P9-BIZ-446

REAL ANIMAL HEROES

Other Titles by Sharp & Dunnigan:

The Congressional Medal of Honor – The Names, The Deeds

The complete reference to the citations for the nation's highest award for heroism, Civil War through Vietnam, with history, photo of medals, and index. ISBN 0-918495-01-6

The Navy Cross – Vietnam

For extraordinary heroism – citations of awards to men of the U.S. Navy and U.S. Marine Corps for the nation's second highest award for heroism, complete with history, photo of the medal, and index. ISBN 0-918495-15-6

REAL ANIMAL HEROES

*True Stories of Courage
Devotion and Sacrifice*

Edited by
Paul Drew Stevens

Sharp & Dunnigan

PUBLICATIONS INCORPORATED

Chico, California

First printing 1988

REAL ANIMAL HEROES, Copyright 1988 by Sharp & Dunnigan Publications, Incorporated. Printed and bound in The United States of America. All rights reserved. No part of this book may be reproduced in any form without permission in writing from the publisher, except by a reviewer who may quote brief passages in a review.

Book designed and edited by Paul Drew Stevens

Library of Congress Catalog Number 86-62119

Library of Congress Cataloging-in-Publication Data

Real animal heroes.

    Includes index.
      1. Domestic animals – Biography. 2. Animals – Biography.
3. Heroes – Biography. I. Sharp & Dunnigan Publications.
II. Title: Animal heroes.
SF76.5.R43 1987   636   86-62119
ISBN 0-918495-25-3

Published by
Sharp & Dunnigan
Publications, Incorporated
165 Piper Avenue
Chico, California 95926

Acknowledgements

To list here everyone who has contributed to this book would require a book in itself. Those intimately connected with its development are mentioned below. Organizations and individuals who have offered records and referrals to further research are listed in the back of the book under *Acknowledgement of Resources*. As with the animal heroes presented, there are many friends who may be omitted from this first volume, but not forgotten. Thank you all.

Harley S. Shane
Paul D. Stevens
Publishers

John Kostelec, Illustrator (Biography Page 9)

Zu Vincent, Principal Writer
Nadine Crenshaw, Contributing Writer
Brian Sobel, Contributing Writer
Dona Gavagan, Contributing Writer

Wilson Dillard, Associate Editor
Beau Westover, Associate Editor

Dona Gavagan, Editorial Assistant
Christopher Stevens, Editorial Assistant

Alex Doyle, Research
Coleen Doyle, Research
Cindy Penrose, Research
Tammy Webb, Research

Mary Devlin, Production Assistant
Jan Haugh, Production Assistant
Debra Woodbeck, Production Assistant

Dave Furgason, Cover Graphics Production
Karen Jackson and Margo McGuire, *Typografx*, Typesetting and Production

Dedication

This book is dedicated to all of those creatures who have given their lives so that humans may live; to Christopher Stevens – who recognized the need for this book; and to Thunder and Lightning, who gave us all their love.

Contents

(Continued)

Contents *(Continued)*

List of Illustrations

About the Illustrator

John Kostelec was born in 1963. One of 11 children, John was inspired and encouraged by his supportive family to pursue his natural talent in art. John attended The Academy of Art in San Francisco and California State University, Chico, graduating with a degree in fine arts in 1987. John's talents cover a broad spectrum of creative styles and media. His works in *Real Animal Heroes* are in the traditional style of story telling: high energy, key moment illustrations that meet the modern reader's demands for precision and detail, yet capture the spirit of early classic story illustration.

Preface

In 1987 a German photographer recorded an event virtually unique to human eyes: a drama on an African river involving a hippopotamus, a crocodile, and a small antelope.

The crocodile, his jaws clamped on one of the antelope's hind legs, was dragging his prey into the river. Suddenly, an enraged hippopotamus charged the crocodile, causing the reptile to release his hold and retreat.

With gentle persistent nudging the huge animal coaxed the wounded, bleeding antelope out of the water and up the embankment. Then, while vultures collected around them, the hippo stood over the fallen creature, nuzzling and protecting. Its massive jaws opened and lightly took the small delicate head of the antelope between, to softly rock it, as it would a calf of its own.

Eventually, the antelope, dead or dying, was left to its fate. The crocodile and the vultures did what they have done for a million years.

But, for those few hours, a record of compassion was made; a record seemingly contrary to the theories of those animal behaviorists who reject the notion that animals can feel compassion – especially one species for another. Behaviorists might describe the actions of the hippo as perhaps those of a cow who had recently lost a calf. Since hippo and crocodile are natural enemies, it was instinct for the hippo to charge. Then, what followed was a playout of substitution.

Perhaps this is so – perhaps the behavior of all of Earth's creatures, including humans, may one day be explained in formulas of cause and effect.

There are those who believe that the term "hero" should apply only to humans. For it is believed that only humans have the ability to reflect, and therefore understand their own mortality, the consequences of their acts, the potential of death. Perhaps.

In the greater perspective, who is to know? Animals demonstrate joy, and they demonstrate grief. Those who have had the pleasure of sharing their lives with friends we call pets, of whatever species, know of a special bond. Animals do not communicate with humans in words. Animals communicate by their acts. They communicate through the devotion (some call that love) and trust that they, in their own way, bestow upon us.

Occasionally we become aware of this devotion when a crisis brings a particular animal into a dramatic and publicized event: an anonymous German shepherd who nuzzles, cleans and warms an abandoned human newborn left to die in an alley, then disappears as help arrives; the cat who calls incessantly until the household awakens, just in time to escape the flames that engulf the family home; the trained companion who, day after day, protects, comforts, serves and loves their disabled and dependent comrades – their humans. They are all heroes.

The stories presented in this book represent the supreme efforts of but a few of those animals who, for reasons of instinct, love, loyalty or sheer endurance, have protected and saved human life – for as long as there has been a bond between mankind and those creatures with which the Earth is shared.

Perhaps from these stories will come a greater understanding of the intelligence, the loyalty and the simple needs of those who give so much for so little: a warm place to sleep, enough to eat – and love.

Patches

While Marvin Scott pried at the stern line of the iced-in patrol boat, Patches narrowed his eyes and lifted his nose into the wind. The frigid blasts sweeping across the dark waters of Lake Spanaway tossed the fur along the white ruff of his neck. The temperature that December night in 1965 had plummeted to nearly zero and, on the small wooden pier near their Washington home, the collie-malemute shivered.

Marvin hadn't expected the dog to accompany him down the rocky, 300-foot slope from the house, thinking the weather would be too miserable for him. But Patches had followed anyway, scrambling along the frozen incline at his master's heels. If Marvin were going to face the bleak night, so would Patches. Marvin was worried about ice damage to the boat, and Patches watched anxiously as he struggled to free the lines.

Marvin leaned out against the ice wedge in his hands and then suddenly tumbled forward, feet slipping helplessly on the icy wood, his arms clutching the air in panic. The startled man's shout ended abruptly with the awful thud of flesh hitting wood as he careened against a floating dock, striking his head and tearing virtually every muscle and tendon in his legs. In another instant his inert form splashed into the freezing void of the harbor.

Anxiously searching the churning waves for the spot where his friend had disappeared, Patches gave one agonized bark, then leaped into the lake. He dove through 15 feet of water, frantically searching about in the blackness. Ignoring the glacial shock, he suddenly glimpsed the thatch of Marvin's hair and clenched his teeth around it.

Marvin outweighed him by more than 100 pounds, but Patches, straining mightily, hauled the dazed man toward the surface. His lungs ached for air. His legs powered stiffly against the numbing cold. Up and up Patches struggled until suddenly, giving a last few urgent kicks, he and Marvin broke the surface with a collective gasp. Stubbornly, his teeth locked into Marvin's hair, Patches powered toward the dock, still 20 feet away.

Patches' breath came harsh and jagged between his clamped jaws. The wind howled across the choppy surface, driving water into his nose and lungs. Beside him the semiconscious Marvin shivered violently.

At last, aching with cold and exhaustion, choking on swallowed water, Patches brought Marvin within reach of the dock. Marvin clutched at the wood and Patches released his hold on him. It was up to Marvin now – Patches couldn't get out of the water alone. He felt himself being pushed to safety by his friend, but once on shore he turned to see Marvin, overcome by shock, slipping under once more. Marvin had blacked out!

Immediately Patches dove back into the lake, once again finding and seizing Marvin by the hair. Again he pulled him to the surface and towed him against the assault of wind and waves to the dock. Once there, Marvin gasped and floundered, fighting to revive himself. Beside him Patches swam in tired circles, waiting, until finally Marvin was able to again shove the dog up onto the dock. His legs useless, Marvin clung weakly to the edge, and Patches could hear his faint cries for help being swept away by the wind.

Marvin was fading fast. If he slipped beneath the surface again he would surely die. Patches paced frantically before the drowning man, licked at the pale, near-frozen fingers and whined. Then, steeling himself against his own exhaustion, he planted his four feet firmly on the rough planks, gripped the collar of Marvin's coat in his teeth, and pulled.

Slowly, Marvin began to respond. With one final effort, he hauled his battered body along the dock and, with Patches still pulling tenaciously at his coat collar, managed to pull himself to safety before collapsing in exhaustion.

Long into that freezing December night, Patches hovered above the soaked and trembling man, listening to his labored breathing. Gradually it began to calm, and eventually Marvin was able to lift his head. Now Patches took hold of his collar again, urging him toward the path to home. Marvin understood. Feebly, he crawled from the dock, dragging his injured legs behind him.

The assent was painfully slow. Both man and dog were near collapse, and the frozen incline rose up as if to torment them. Patches was shaking uncontrollably from the cold, his spent muscles cramping, but Marvin was almost gone and so, pulling at Marvin's collar with all his strength, Patches strained up every bruising inch of the 300-foot path to the house. After an eternity of effort, Marvin at last was close enough to the back door to throw stones against it and alert his wife.

For 25 days after Patches rescued him, Marvin Scott lay dangerously close to death. Even then his recovery was slow, and it was six months before he could return to work. Clearly, he would not have survived the ordeal if the courageous Patches hadn't tagged along with him down to the docks that December night. Others thought so too. For his incredible heroics, Patches was named the Ken-L Ration Dog Hero of the Year for 1965.

Patches

Shade McCorkle

Shade McCorkle, named for his dark, gray striped coat, purred softly as he nestled into the afghan of the same color tucked across his mistress' feet on the day bed.

Broad-shouldered and short-legged, Shade was a cat of mixed pedigree and unknown origin. He had joined the household three years before, distinguishing himself by refusing to eat anything but salmon, beef and milk, and by developing a close attachment to the frail, 88-pound Nell Mitchell, whom Shade accepted as his mistress. On days like today, when she lay ill, his closeness filled Nell with a warmth far more comforting than any blanket.

It was late December in 1932. Mrs. Mitchell's housekeeper had summoned Doctor Longset and left the Memphis, Tennessee, woman to his care. The doctor gave her an injection of medication, listened to her reassurance that her husband would soon be home, and went on his way. Alone, Nell drifted in and out of a fitful sleep.

There was a knock on the porch door, but neither she nor Shade stirred. Unable to rouse herself, Nell hoped whoever it was would simply give up and go away. Instead, she heard the screen door of the porch being pulled open and someone walking toward the entrance to the kitchen.

Again the knocking, but this time closer and from inside the house. Nell lay still, helpless to answer. At her feet, partially covered in the afghan, Shade lazily rolled over. Nell was startled when the kitchen's inner door swung open.

Now the intruder was in the room, a thin, grubby-looking man in his early twenties. His dark eyes darted around the room, taking in Nell's situation at a glance. Belligerently he asked her for something to eat.

"Go next door," Nell winced, "I'm too ill."

Instead, the stranger grinned and pulled up a chair. He settled into it with a menacing nonchalance.

"Who was the tall guy who just left?" he demanded suddenly.

Nell tried to look him in the eye, but the medicine had made her groggy. "Dr. Longset," she managed to reply, her tongue thickened by the drugs.

"When will your old man be back?"

Nell was confused. She thought that, perhaps, this was some strange joke. Beneath the covers Shade lay, forgotten and still.

But it wasn't a joke. Suddenly the intruder was out of the chair and standing over her.

"Your ring!" he shouted.

Shakily, the frightened woman pulled the ring from her finger. She was weary enough to hope the man would just rob her and go away.

He didn't. In a fit of unprovoked anger, he lifted her roughly by the collar, pulled back his open hand and slammed it against her cheek. The slap resounded through the empty room.

The intruder never got another chance at her. Shade's 14 pounds of feline claws, teeth and muscle sprang from beneath the afghan in a hurtling fury.

Before the stranger could see what was attacking him, Shade bounded onto his shoulders. In a flash the cat circled to the man's throat and, with a fearsome yowl, clamped down hard with his fangs. His claws continued to flail in their relentless attack.

The man hollered out in pain as bloody scratches appeared on his face and neck. He whirled around the room, punching at Shade in an effort to dislodge him, but Shade hung on.

The man stumbled into a table, knocking a lamp to the floor. He tripped over a footstool and bashed his hip against a chair. Blood dripped onto Nell's carpet from the fang and claw wounds. The man's hands found Shade's neck and he tried to throttle the life out of the fearless feline.

"Call him off!" the man screamed at Nell.

Shade, however, was far from finished. He squirmed out of the man's suffocating grip and again lashed out at the intruder's face, sending skin and blood flying. The intruder, outweighing Shade by at least ten-to-one, backhanded him unmercifully, but Shade didn't falter. Instead, he set upon the man with renewed vigor, climbing and clawing until the now-terrified attacker screamed in agony and backed toward Nell's front door.

As he retreated, the man got another choke hold on Shade and, upon reaching the door, viciously slammed the enraged cat against a table. He'd had enough, however, and he dashed out the door before Shade could recover and launch a new attack. The last thing Nell heard before she fainted was the man's footsteps retreating pell-mell down the street.

Shade, although bruised and shaken, suffered no lasting effects from the desperate battle with the intruder. And, for protecting his mistress from an unknown fate, the heroic and determined cat was awarded the Latham Foundation's Gold Medal for animal heroism.

Buster

All 35 people in the Minneapolis, Minnesota, apartment building slept soundly, unaware of the fire. Buster, a small, black-and-tan Spitz belonging to Mr. and Mrs. Frank Remackel, woke slowly at 4 a.m. on April 13, 1930, the day of the fire. Dense smoke had already obscured the rooms of the Remackels' flat.

Buster slept with Fluffy, the Remackels' Persian cat, who'd been his constant companion since the day he'd saved her life the year before. Then only a few months old, Fluffy had gotten caught in a small trap and lay unable to move or escape. It was Buster who found her and carried her home to Mrs. Remackel, trap and all.

Now he woke the sleeping cat. As Fluffy rose, Buster left their kitchen rug and poked his head through the door to Mrs. Remackel's bedroom. His mistress lay unmoving under the covers. Crossing the carpet, the dog pressed his nose to the pillow and started licking her cheek. Groggily, she pushed him aside.

Buster whined and jumped up on the bed. When his mistress rolled away from him in annoyance, he tugged at the covers, barking. Mrs. Remackel, unaware that the suffocating smoke was lulling her into unconsciousness, pulled the pillow over her head like a child. At this Buster yelped shrilly, settled his teeth around her arm, and clamped down. The bite did not break the skin, but it was enough to make Mrs. Remackel sit up in alarm. Awake, she found she could hardly breath through the smoke.

Confused and frightened, Mrs. Remackel groped toward the second bedroom, her mind filled with thoughts of her husband, who slept there. The walls reeled when she stood and she grapped for a handhold to steady herself. She was panting with the effort of each step.

While she faltered, Buster yelped and bounded toward the main door. Without hesitating, he crashed into its hard wood surface, barking wildly. He was still flinging himself into the wood when the dizzy Mrs. Remackel reached it and let him out.

Free of the room, Buster raced along the hallway to the first apartment door he came to. Throwing himself against it as he had Mrs. Remackel's door, he barked at the top of his lungs. Soon a drowsy tenant was staring down on him from the threshold.

Buster didn't wait. He was at the next door instantly, shouldering into it and barking shrilly. As soon as he heard an answering commotion within, he moved to another and barked and scratched.

All along the long hallway doors began to slam open as Buster crashed against them, sounding the alarm. At the last door Buster paused, whining pitifully. There was no answering shout from within. The smoke was billowing into the halls now, and the residents were running up and down half-clad, bumping into one another and hacking out instructions through painfully seared lungs.

Buster studied the closed door. Suddenly the 25-pound dog flung himself against the unyielding wood with such force that the latch gave, and he landed in a heap inside. Buster shook himself and bounded up, racing into the bedroom where an elderly man, an amputee with an artificial leg, lay sleeping. Springing onto the man's bed, the dog yelped and growled, pulling at the bedcovers until the man at last sat up.

The man blinked into the smoke and made a wincing, slow effort to rise. Buster scuttled in circles and barked, dancing around the narrow bed but not leaving it. The amputee, who walked with a cane, had to perform a series of halting movements, attaching his limb and pulling on an overcoat and slippers. Smoke was seething through the rooms, and the heat of the fire was growing intense, but still Buster stayed with him as he fumbled at his tasks. Finally the amputee was ready, and Buster followed him into the corridor and down the steps leading to the street.

Mrs. Remackel, having reached the stairs with her husband and several others, called to Buster as she saw him descend. The Spitz, however, turned back once the amputee was safely down. In their haste and confusion, the Remackels had forgotten Fluffy. Buster gave his mistress one agonized glance and disappeared into the smoke.

When he re-entered the building, the smoke was now taking its toll. His normally agile run slowed to a a half-trot as he wove back along the murky hallway. His eyes were thick with tears. After a long, ungainly trek he neared the door to the Remackels' apartment. Edging inside, he paced the smoldering rooms. There was a rush as the smoke gave way to flame; a hot racing inferno that would completely destroy the building. The exhausted Buster sounded a single bark.

When Fluffy appeared out of the haze, Buster leaped at her, chasing her toward the door. Rushing behind the frightened young cat, he nosed her into the hall. Then, herding her like a skittish goat, he nipped at her heels to get her across the blistering floor. As the fire roared behind them, the pair made the stairs and were gathered up by the waiting Remackels. Buster had saved Fluffy – and 35 human lives.

No one was injured in the fire that gutted the Minneapolis apartment building. Buster's quick-witted heroics so amazed his city, along with animal enthusiasts across the country, that he was awarded the Latham Foundation's Gold Medal in 1932.

Old Soup

The line of elephants moved rhythmically along the dock near Cawnpore, India, in the summer of 1937. The early morning air, bronze with dust and rising heat, steamed against the cargo bags of rice each elephant carried to the waiting barge. Beyond, the broad Ganges River glinted in the sun.

First in line, Old Soup stepped delicately despite his lumbering side-to-side gait. His ancient, parchment-like ears flapped as he heard the shouts of Major Daly's soldiers while they supervised the loading. It wasn't the soldiers, but the children, who interested Old Soup. Owned by Daly, he was delighted by the familiar sight of the officer's young son and daughter who had come to watch the procession on the dock.

With the youngsters in front of him, Old Soup had to raise his great trunk and look down his nose to see them. But as he passed, he could get a better view with his elephant's curious backward vision, and he trailed his sensitive trunk along the ground while he watched. His gaze took in the boy and girl waving back at him as they stood on either side of their nursemaid.

Old Soup halted on command and stood blinking his long lashes in the sun. His lips folded over massive tusks. The right one was slightly larger than the left, and both were elegantly curved and yellow with age. Old Soup was said to be 100 years old.

As Old Soup was relieved of his rice sacks, there came an angry blast from one of the other elephants. Suddenly gone mad, the five-ton beast broke from the line and wheeled on his keeper. He lifted one mighty foot and slammed it down on the surprised man, crushing him instantly. Turning again, he smashed brutally against the side of the barge, breaking open the bags on his back and flinging rice into the Ganges like confetti. His huge trunk unrolled skyward and trumpeted defiantly. Then, ears back, trunk drawn down against his chest, the enraged elephant thundered straight for the Major's children.

As the mad elephant broke from the line, the children's nursemaid, in panic, began to pull them toward the safety of a building. She looked up again in horror to see the enormous beast bearing down, wild eyes sighting them over the spear-like tips of his tusks. She froze.

At that moment Old Soup pulled away from his handler and trumpeted fiercely. He lowered his own great trunk, flapped his ears to his sides and raced for the demented pachyderm. Too far away, Major Daly and his soldiers could only watch helplessly. The children, their escape route cut off, were separated from certain death only by Old Soup's determined charge.

Just a few feet from the children the elephants collided with the force of crashing locomotives. The thunder of their clash shook the earth. In a riot of dust both elephants rebounded. A huge welt raised on Old Soup's forehead. Nearby, the children huddled in terror.

Once again the rogue elephant turned on the tiny humans and raised his gargantuan foot, the blood of his handler barely dry on his skin. The nursemaid screamed. Old Soup barreled forward, slamming the enormous brute backwards. The colossus sounded his fury and started for the children again.

Old Soup blared and lunged forward, diverting his adversary's attention. The ground trembled and the air vibrated as the two gray mountains locked tusks. Head to head, both elephants pushed, Old Soup venting his rage in another trumpet blast. The clatter of their tusks echoed like gunshots along the quay.

The rogue pulled up swiftly and jabbed, catching Old Soup's ear on the point of his tusk and ripping it nearly in two. Old Soup retaliated by grabbing his attacker's trunk in his own and twisting violently. Younger and more agile, the other pachyderm squirmed free.

Major Daly and his soldiers made frantic attempts to reach the children trapped in the dusty confusion, but any new movement further enraged the crazed bull.

Bruised and battered, both ears in tatters, Old Soup at last caught his opponent off guard. Twisting his mighty head, he stabbed his tusks into his opponent's vulnerable underside. Once, twice, Old Soup thrust and withdrew. The younger elephant staggered, blood from his wound zig-zagged along his leathery hide. But then he was at Old Soup with renewed rancor, rushing wildly and plunging his tusks with abandon. With a sickening crack, Old Soup's right tusk splintered halfway to his gums.

The earth trembled and thunderous trumpeting rent the air as first one and then the other elephant plunged their tusks into flesh, wrenched upward, and withdrew. Elephants rarely fight to the death, but this was going to be an exception. Helpless to intervene, the onlookers watched in disbelief.

For more than an hour the battle raged. Huddled and helpless, the children and their nursemaid watched with dirty, tear-streaked faces. Finally it was over. One elephant's tusks hit home, sinking into the other's vital organs. Fatally wounded, the defeated giant staggered to his knees and rolled over, dead.

In the settling dust the victor stood blinking before the torn and bloody carcass, then raised his mammoth trunk in a long, eerie blast. Blood oozed from the stab wounds in his chest, his right tusk was shattered, his tattered ears hung limp, but Old Soup had won his fight to save the children. Shaken but unharmed, they squirmed from their nursemaid's grasp and rushed over to the brave beast. Gently, Old Soup reached out with his trunk and nuzzled them.

Old Soup recovered from his wounds. After his magnificent fight, Major Daly retired him from service and took him home for a pet. The protective pachyderm became the best nursemaid the children ever had, and could often be seen taking them for a stroll in the countryside.

Old Soup

Leo

The sky over Hunt, Texas, was so blue that specks of sunlight danced in Leo's eyes on that August day in 1984. The four-year-old white royal standard poodle was standing up in the jeep, grinning into the hot wind. He was on his way to Honey Creek, near the Guadalupe River, with his master, William "Bud" Callahan; the two Callahan children, Sean, eleven, and Erin, nine; and Boo, the family's big black Labrador.

Bud braked the jeep near the driftwood-littered curve of the creekbed. Right away the children began searching out sticks for the dogs. Sean tossed one for Leo, yelling, "Go get it, boy!"

Downstream, among the roots of a big cypress tree, a diamondback rattlesnake lay, waiting out the worst of the day's heat. The rattler was big — five-and-a-half feet long. Though deaf, like all snakes, this one knew by the vibrations that his secluded hideaway had been invaded.

The snake lifted his head. His slick scales, overlapping like roof tiles, rippled. His forked tongue flicked out, picking up the scent of warm-blooded prey. And now he paused, still as a spider in its web.

Meanwhile, Leo had thrown his 50 pounds into the waters of Honey Creek and was returning to the bank with the stick Sean had tossed. The boy was running downstream, on the lookout for another good stick. Erin ran close behind him. Leo angled through the water toward the boys on the bank.

Just before Leo hit land, Sean screamed. At the sound of the boy's sharp, thin voice, Leo dropped the stick and came out of the water, all his protective instincts bristling.

Sean had knelt among the gnarled roots of the cypress and there had come face-to-face with the unblinking diamondback. The snake, cornered, wasn't shaking his loose horny tail in warning; he was already coiled, his head reared and weaving, about to strike at Sean's head or chest.

Leo didn't hesitate. Barking furiously, he lunged, placing his body between Sean and the deadly pit viper. The snake's mouth opened nearly flat, his fangs folded out, erect, glistening momentarily, and then he struck.

The two long, hollow teeth felt like drill bits in Leo's skull. He fell back from the force of the blow, with the heavy-bodied snake following, broad belly scales and strong muscles moving with lethal grace. White fangs stabbed into Leo's head a second time, withdrawing immediately as the snake slid back to rearrange his coils.

Leo went in for his own attack now. His head felt oddly numb, yet he dove for those pale-bordered diamonds undulating before him. But the snake's defensive coil protected him as he pitched his head and fangs at the dog a third time.

25

Now Leo stumbled. He felt dazed. The snake's venom was already working on his central nervous system to produce eventual respiratory failure and suppress the action of his heart. He only vaguely heard Sean and Erin screaming for Bud. He saw more surely the snake's reptilian stare weaving atop the stalk of its neck.

Still the children's hysteria penetrated his consciousness enough to stir him to keep trying to drive off this hissing threat. Growling, lips lifted to show his own teeth, Leo jabbed in, trying to seize that slim, taunting neck.

The snake flowed smoothly forward to meet the dog.

Leo was nearly immobilized. He hardly had time to close his eyes as the fangs came blurring at him yet again to pierce the left side of his face.

Bud Callahan had heard Leo's barking and the children's fusillade of cries. Watching the battle between Leo and the snake out of the corner of his eye, he herded his terrified children to the safety of the jeep. Boo, the big Labrador, had already high-tailed it, wisely if not courageously.

Leo refused to give ground. Once more he rose to attack, but he found he couldn't move. He couldn't do anything but watch the diamondback slide off. The venom had taken its toll.

Bud returned to the cypress tree to find Leo lying alone among the roots. He lifted the dog's bloody body and made another dash for the jeep.

Back at Honey Creek Ranch, Leo's mistress, Lana Callahan, began making repeated but futile calls for a veterinarian. At last she reached one in Kerrville, 14 miles away.

Now the whole family piled into the jeep. Already 45 minutes had elapsed since Leo's battle. His head and neck were swelling grotesquely from the huge amount of poison that had been introduced into his system. His collar was choking him. Lana cut it off, allowing him to suck in a little more air.

By the time they reached Dr. William Hoegemeyer's clinic there was such gross swelling of his head that he was unrecognizable. Hoegemeyer's prognosis was honest, and guarded in the extreme.

"There is so much trauma here," the veterinarian told the distraught Callahans, "and the damage to the tissues is so extensive, well... I doubt he'll make it."

He knew young Sean could never have survived such a large dose of heart-depressing venom punched directly into his chest. And yet – miraculously – Leo was alive. Hoegemeyer administered two bottles of antivenin and prepared to watch the dog around the clock.

Hours passed, and then days. Leo's fiesty heart pumped on. Not only did he live, but he recovered with no dreaded brain damage – without even any loss of vision in his endangered left eye.

The heroic dog was his grinning self again when he left his suite at the New York Hilton Hotel in May of 1985. He trotted into the studio of television's *Good Morning, America* and there became the 31st recipient of the Ken-L Ration Dog Hero of the Year Award. He was the first poodle to receive it.

In 1986, the Texas Veterinary Medical Association inducted the lion-hearted Leo into the Texas Pet Hall of Fame.

Mac

Early in the year of 1982, a dirty, thin and nameless German shepherd wandered into the Police K-9 Training Center in Trenton, New Jersey. To the training officer on duty, Sergeant Pete Schwendt, it was as though the weary loner had come to enlist.

Beneath the dog's scrawny, neglected exterior, the sergeant thought he saw something – some special, unexplainable quality hidden deep within the dog's brown eyes. He named him Mac, and the homeless shepherd was fed, groomed and assigned to K-9 school with his new partner, Officer Robert Parrish.

By the end of the 14-week training period, a fondness had developed between Parrish and the shepherd, and the important K-9 qualities detected by Sergeant Schwendt bloomed in Mac. Not the least of these was his affection for Parrish. Working side by side during the months from June to September, their mutual trust and affection deepened. During this time, Mac was responsible for four suspect apprehensions, receiving his first Citation of Merit in July.

On a routine patrol the morning of October 14, Officer Parrish and Mac listened to a scratchy, pre-dawn call. Police summoned to a burglary in progress were requesting K-9 assistance at a tavern in downtown Trenton. They believed they had the thief or thieves still trapped inside, and it would be up to the K-9 team to make a search.

"That's us," Parrish nodded toward his partner, turning on his lights and weaving the patrol car through traffic. It was 2:34 a.m.

In the pale city darkness that passed for night, Mac stood tense beside Parrish on the street outside the tavern, his fur washed with the surrealistic flash of squad car lights. Behind them a small crowd of curious civilians gathered. Parrish lifted a bullhorn and commanded those inside to surrender. The onlookers shuffled expectantly. Parrish and Mac waited.

At last Parrish gave the signal, and Mac moved cautiously ahead of him toward the silent tavern. Parrish unlocked the barred front door with the owner's key and the K-9 team slipped into the deeper blackness.

As they entered the tavern, Mac hunkered forward, ears pricked upright and nose to the ground. Behind him Parrish searched for the lights. When he flicked them on, shadows rose in long fingers cast from the legs of chairs overturned on tabletops. The room was empty.

Mac didn't hesitate. Loping forward, he led Parrish through the deserted room, halting in front of a closed door. Here the shepherd gave a low growl and the hair along the back of his neck went stiff.

Parrish gestured for silence and turned the knob, crouching aside to take momentary cover. There was no attack. Mac stepped into the opening. A dark stairwell fell away before him, while light cast from the room behind revealed the wall switch several feet below. He took the lead as he and his partner descended toward it.

With each step the maw of the dark cellar remained eerily quiet. Although Mac's fur stayed taut, it began to appear as if they'd find this room deserted as well. Finally Parrish reached the switch and turned it on.

Just as the room flooded with light there came a sound off to his right. Before he could turn, a man leaped from hiding behind a wall, a huge hunting knife gripped in his fist. He lifted the blade barely five feet from Parrish's back. In the split second he brought it down, Mac sprang.

Lunging high, Mac threw himself between Parrish and the assailant, taking the full force of the knife meant for Parrish. The blade tore the big police dog's side wide open. Even as it did, Mac spun and clamped his teeth around the assailant's weapon arm.

Blood spewed wildly and Mac fought to keep his grip. Growling, he held on for a stubborn moment, his attacker whirling and punching at him with his free arm. Mac's body was contorting in pain; his strong jaw muscles slackening with shock. His vision blurred and he fell, snapping at the thief's legs as he went down.

Because of Mac's intervention, Parrish had been given a few precious seconds, and now he quickly disarmed the thief. Revolver out, he motioned his assailant away from Mac. Within minutes other officers came to their aid and Parrish, relieved of his duty, gathered the bleeding shepherd gently into his arms and rushed him from the tavern.

Back in the squad car, he sped the now-unconscious dog to a nearby 24-hour veterinary clinic, but Mac's wounds proved too serious. It was 3:26 a.m. In less than an hour after they'd responded to the call, Mac was dead. He'd taken the knife thrust meant for Parrish – and given his life to save his partner.

Mac was the first dog in the 21-year history of the Trenton Police K-9 unit to lose his life in the line of duty. His death occurred just one week before he had been scheduled to stand formation with Officer Parrish to receive a second citation for outstanding K-9 performance.

Instead, that citation was awarded posthumously, along with a second posthumous award – the Citation for Valor – during a burial service with full honors conducted for the courageous dog who saved Parrish's life.

Mimi

Mimi had been sleeping on the rug. Struggling awake, the miniature poodle peered around the hazy living room she knew as home. Something was terribly wrong. The air was thick and choking, with a smell like burning rubber. There was an ominous noise – the noise of a vast flight of birds, their whooshing wings accompanying the grim smell.

Always dainty, Mimi sprang onto her manicured nails and shook her puffy ears in dismay. When she popped up, the thin collar at her neck jangled softly. Her owner, Nicholas Emerito, was slumped forward on the sofa in a deep sleep. Before him hissed the empty TV screen.

It was nearing 5 a.m., several hours after Nicholas had fallen asleep watching the late show in his Danbury, Connecticut, home. As fire swept through the two-story house that morning of January 30, 1972, he, his wife, and six children slept unaware.

Instantly, Mimi leaped onto Nicholas' chest and scratched at his shirt. She barked sharply into his sleeping face. With smoke in his lungs, Nicholas coughed, muttered and cuffed groggily at his dog, the lethal vapors urging him back to sleep. Mimi barked again, a high-pitched yapping that intruded roughly into Nicholas' murky brain.

Suddenly he sat up and gripped the little dog tightly, blinking into the smoke. The room was in flames! Nicholas reeled to his feet, his forearm against his face to shield it from the intense heat. The fire, which had started in the kitchen, had set the whole house ablaze. His mind raced urgently. His wife and five-year-old son, Peter, slept downstairs, but his five older children were in bedrooms on the second story. In those first panicky seconds he rejected the fear that there might not be time to warn them all.

Nicholas stumbled up and ran toward the downstairs bedrooms, shouting hoarsely as he went, but his warning was lost in the roar of the flames. Dimly he was aware of Mimi, racing ahead of him toward the stairwell.

Skipping across the burning floor, yipping as she ran, Mimi flung herself at the staircase and scrambled up its hot treads. In a matter of moments fire was everywhere. It licked up the stairwell after the little dog, and whirled in furious plumes across the ceiling.

Mimi rushed madly into the girls' bedroom and tugged at 11-year-old Lisa's bedcovers. When Lisa cried out sleepily, Mimi hopped up on her chest and scratched insistently. She tugged on her pajamaed arm. Finally Lisa sat up.

Mimi darted across to nine-year-old Patty and began the same frantic bid for attention. Scrambling atop her huddled form, Mimi yapped continually and drug her claws over Patty's back.

When both girls had tumbled bewildered from bed, Mimi turned and skidded down the burning stairs to Nicholas, who had managed by this time to rouse his wife and Peter.

As Mrs. Emerito left the house with Peter, Nicholas hurried through the smoke-blanketed rooms with Mimi, dodging the flaming paint which curled and rained from the ceiling. Lisa and Patty were crouched halfway down the blazing stairs, faces blanched with fear. Quickly Mimi ran to them and pulled at their nightclothes, urging them down the last few steps into their father's arms.

Flames whooshed up the stairwell, making descent for the other children impossible. Nicholas shouted and grabbed at Mimi's collar, trying to get his dog and the shaken youngsters from the house. But Mimi bounded out of his reach and refused to follow. Head high, she leaped up the gauntlet of flaming stairs, while the inferno closed in behind her.

Upstairs, Mimi burst into Debbie Emerito's room, snarling maniacally at the 15-year-old. Her little pink tongue hung like a dry petal as she darted back and forth. Groggy from smoke, Debbie fought Mimi's attempts to rouse her, but at last the girl woke.

Next, Mimi raced for Anthony, 13; and his brother, 16-year-old Edward. Whipping Anthony's covers from him with a fierce side-to-side tug of her little teeth, she nipped at the boy's head. After a few anxious moments he shoved her aside and sat up, rubbing his eyes.

By now Nicholas had climbed a ladder positioned outside his sons' bedroom. Mimi barked at him through the window and raced to wake Edward, performing her passionate dance one last time. As soon as Edward was alerted, she began to herd the three older children toward the waiting Nicholas, alternately circling their feet and tearing ahead of them to the glass.

Lifting the window, Debbie stepped onto the ladder and called to Mimi, but the little dog ducked behind Edward and stood her ground, again refusing to leave. Debbie had no choice but to descend the ladder hastily behind her father. But as Anthony tried to go down, fire swept across his path from below, igniting their escape route. The ladder was now useless.

"The roof!" Nicholas shouted from the ground, gesturing for the boys to climb out and jump. Edward took Mimi in his arms and, with Anthony following, crawled out onto the shingles. Flames from the inferno in the rooms beneath their feet sparked across the sky. There was the sound of sirens in the distance. The trapped boys looked down at their father, who pleaded with them to jump.

"You first," Edward ordered his younger brother. Mimi struggled from his embrace and barked encouragement. Anthony stared at Nicholas below and gathered his courage. Then he leaped into his father's arms.

Once Anthony was safe, Edward reached for Mimi. But the exhausted poodle barked a warning and danced up the steep slope out of his grasp. She did not intend to leave Edward behind. Long red fingers were curling toward the space where they stood, and the boy was forced to jump first, hoping that Mimi would follow.

Mimi

Nicholas held out his arms again and broke Edward's fall. Then he turned back to Mimi. The entire Emerito family was safe now; safe and waiting. Satisfied, Mimi gave a bark, gathered her small legs beneath her, and sailed off the blistering roof.

The early morning fire completely gutted the Emerito home. Fire Chief Joseph Berialowitz, arriving on the scene shortly after 5 a.m., praised Mimi's courageous actions in warning the Emeritos in time. If not for her, he noted, eight people might have died.

In recognition of that fact, Mimi was presented with the Ken-L Ration Dog Hero of the Year Award for 1972. The frisky little female was the first miniature poodle ever to receive the award.

Villa

The storm made Villa restless. The shaggy Newfoundland, named after the town of Villas, New Jersey, where she lived, paced back and forth in front of the kitchen door to be let out into her run. Her owners, Dick and Lynda Veit, had just taken her and their other two Newfoundlands for a walk on the beach, but the onslaught of a fierce storm off the Cape May Peninsula and a bitterly cold temperature of 30 degrees Fahrenheit had caused them to cut it short.

The angry nor'easter that blew in that Friday morning of February 11 would later be called the "Blizzard of '83," but for now it was just another storm to the broad-muzzled dog. At a little over a year old and already weighing 100 pounds, her dark eyes were patiently attentive as Dick opened the door.

Behind her, the other two dogs slept peacefully, curled up by the fire and secure in the close warmth of the house. Ahead, the tearing wind hit as powerfully as a giant's hand, bringing with it great gusts of stinging sleet and snow.

Once outside, Villa encountered others who had been restless, too. Through the swirl of white she could barely make out the three Anderson girls, who lived not more than 50 feet away, playing in front of their house next door. Excitedly the trio, which included Villa's special friend Andrea, age 11, were digging tunnels through the great drifts and tossing snowballs at imaginary targets.

Villa stood in her run, a dark shape in the darkening world, still warm in her dense coat of fur, listening. It was Andrea and her sisters who had helped make the long, five-foot-high dog run that was her home, and who fed and petted the Newfoundlands whenever the Veits went away for a weekend. Villa's tail wagged softly back and forth at the sound of the girl's muffled laughter.

It didn't last long. Even bundled up, the children were no match for a wind-chill factor well below zero. "I'm freezing," the eldest Anderson girl declared, taking her youngest sister inside. Left alone, Andrea played for a few more minutes before admitting her sister had been right. It was indeed freezing. She turned in the direction of the house, but she never made it to her door.

Suddenly a slap of wind hit the five-foot, 65-pound girl with such force she was lifted off her feet. Andrea struggled helplessly as she felt herself being knocked brutally backwards. Over and over she tumbled in the fuzzy, frozen world, down a five-foot embankment and 40 feet toward the dunes. She was just yards from the bay when she came to a halt in a deep snowdrift.

She was buried chest deep. Sand and snow obliterated the world with a savage vengeance. Her arms and legs were trapped. Unable to move, Andrea began to scream for help.

In her run, Villa perked her ears forward and nudged her nose against the wire. Was there a sound above the moaning of the wind? A faint wail? A wail whose pitch she recognized?

Buried in the snow, Andrea worked desperately to claw herself free. But she was so tired already, and the harder she worked, the tighter the snow seemed to grip her. The cold numbed her hands and feet into useless clubs. Her throat ached from shouting.

Villa was a squarely built dog, with huge paws and heavy chest. She'd never jumped her pen before. She'd never even tried. But wasn't that little Andrea out there now, crying for help where no one could hear?

Villa was over the fence in a single powerful leap. She landed on her feet and homed in on the weakening cries coming from the direction of the bay. If she could find Andrea it would be her instincts that guided her, for there was nothing to see but swirling snow.

Quickly she trundled over the embankment in search of the frightened girl. Away from the shelter of the houses the gale-force wind was incredible. In places it worked snow into huge drifts, and in other places left the ground swept bare. Villa put her head down and hurried through the storm.

The Newfoundland was almost upon Andrea before the dog saw the bit of parka sticking up through the snow. Andrea gave a small cry and the dog was beside her instantly, plowing through the snow to lick her face.

The little girl was relieved simply by having the dog there. Villa, fighting the increasing fury of the storm, methodically began tramping around and around the buried girl, packing the snow down with her paws. At last Villa stood still with her head toward Andrea, waiting for the girl to grab her neck.

Villa didn't move until the child's arms were tight around her and then, with all her strength, she began to pull. Slowly, slowly, she worked to free Andrea's trapped body from the snow. The small arms clung to her trustingly; the numb legs began to work loose.

All at once Villa was dragging Andrea, belly down, onto a spot of level sand. She stopped and nudged the girl to her feet and waited until Andrea wove her hands tightly into her fur before leading her along the beach toward a more protected draw.

Villa was tiring from her efforts, but Andrea was exhausted and an easy prey to the dangers of hypothermia. Their only hope was to climb the embankment and reach shelter. Carefully, Villa furrowed a path for them with her chest and paws, and they struggled to the top. But there the wind hit them with all its fury. Crying out, Andrea lost her grip and fell backwards.

Once again Villa circled her, taking the brunt of the wind and holding her neck steady for the small arms to grip. Wearily, Andrea struggled up and they were on their way. Again the wind buffeted the huddled pair and again Andrea lost her grip. But Villa didn't waver. In a maneuver Andrea was beginning to understand well, the dog circled her and paused, neck outstretched.

They would repeat this several times in the 15 minutes it took them to cover the 40 feet to the Andersons' house. Panting heavily from the effort, Villa led the girl right onto the porch and scratched urgently at the door. Not until she heard Mrs. Anderson's footsteps approaching from the other side did the now-exhausted Newfoundland slip away and trudge the last 50 feet through the blizzard to her own home.

At the front entrance to the Veits' house, she pushed against the door to rattle the latch. When an astounded Lynda came to investigate, Villa moved heavily inside, tongue lolling, and flopped with the other two Newfoundlands by the fire.

Eleven-year-old Andrea suffered no lasting ill effects from her half-hour of being trapped in the blizzard on Delaware Bay. But, given the wind-chill factor, she wouldn't have survived much longer on her own. The story of how the affectionate Newfoundland saved Andrea's life won Villa the prestigious Ken-L Ration Dog Hero Award of 1983, and the Newfoundland Club of America honored her as "Heroic Newfoundland of 1983."

Lassie

Young Gary Gustafson had insisted on picking her himself from the litter of Shetland sheepdogs. He named the delicate-faced, fox-eared pup Lassie, and she followed him everywhere. At night she slept at the foot of his bed.

That evening in 1956, six-year-old Gary had just come back to the Gustafsons' San Carlos, California, home following a recent tonsillectomy. He'd gone to sleep early, and Lassie had followed him to bed in her usual way. She was curled near him when Mr. and Mrs. Gustafson checked on the pair before going to sleep themselves.

The Gustafsons' room was at the opposite end of the house from Gary's so it was only Lassie who woke later when the little boy started to tremble. She came immediately alert and listened to his gurgled breathing. A victim of sudden hemorrhage, Gary had begun to bleed profusely. He could literally bleed to death in a matter of minutes without medical attention.

Half-conscious, the little boy called out and stumbled from bed in an attempt to reach the door. He didn't make it, collapsing on the floor instead. There Lassie ran to his side, whining and licking his cheek. She was unable to revive him. Gary only stirred and moaned, a dark, wet stream of blood oozing from the corner of his mouth.

Lassie left Gary and ran down the hall to his parents' closed door. Smart and obedient, she'd been taught never to enter their room. Yet for once she ignored her training and threw herself at the door until it swung open. Then she raced into the room yelping and whining.

The elder Gustafsons aroused from a sound sleep to find the little dog running around their bed. While Gary lay helpless down the hall, she worked to gain their attention. Quickly she grabbed the bedclothes in her mouth and yanked as hard as she could.

Mr. Gustafson was confused. Lassie had never before broken the rule about entering their room, and he ordered her out sternly. The dog didn't listen. She only barked louder and shook the bedclothes furiously with her teeth.

Mr. Gustafson sighed and climbed out of bed. He decided she wanted outside and headed toward the back door. Lassie wouldn't follow.

Gary's life was slipping away with each second that passed. Anxiously, Lassie planted herself in the hallway leading to his bedroom door. Then she raced up to Mr. Gustafson and turned, racing back to Gary's room. When he stood there, uncomprehending, she repeated the move.

Twice, and then a third time, with every muscle in her body tense with expression, she raced back and forth in a form of exaggerated pantomime.

Whining, barking, racing back and forth, she used every method at her disposal to convey the message that something was terribly wrong behind that bedroom door.

At last Mr. Gustafson gave a start and, realizing what the little Shetland sheepdog wanted, followed her into Gary's room. There he found his son lying in a pool of blood. It was almost too late.

Gary was rushed to the hospital and given emergency treatment. He'd lost so much blood that there was no doubt in the minds of the attending physicians that Lassie had saved his life. Another 15 minutes would have proven fatal to the little boy.

For saving his master's life, Gary Gustafson's quick-witted and devoted dog, Lassie, was awarded the Ken-L Ration Dog Hero of the Year Award for 1956.

Reckless

Alone, the small sorrel pony galloped through the stubble of the rice paddy heading toward the ridge beyond. The incoming artillery and mortar fire rained around her, on the paddies and slopes of the ridge, at the unparalleled rate of 500 rounds per minute. Strapped in the pony's saddle pack were 192 pounds of ammunition for the 75mm recoilless anti-tank rifles working desperately on the front lines.

She moved quickly with the heavy load across the paddy's rugged and uneven ground. Rivulets of sweat sliced down her haunches and streaked her crusty belly while the howl of mortar shells and the explosions of artillery blasted the sound of her own heavy breathing from her ears.

The nickname for the Marines' unit was "The Reckless Rifles," and so they had named the sorrel, "Reckless," but she was surefooted as she raced across the rice paddy on swift, strong legs. Her destination, Hill 120, loomed 1,800 yards ahead. Positioned there were the 75mm recoilless rifles she serviced.

Suddenly the shriek of a mortar shell split the air too close. Reckless veered but kept going as it came screaming down. A mound of mushrooming earth shuddered and exploded in a flash as the shell struck. Dirt and shrapnel pummeled her and she flinched as hot metal sliced her eyelid and tore into her flank.

Under the heaviest fire attack ever placed on a sector by the Communist Chinese during the Korean Conflict, members of the Second Battalion, 5th Marines, were putting up a bloody fight that day in March of 1953. The stakes were a vital area of the Panmunjon-Bunker Hill Sector known as the Nevada Complex. Outpost Reno had already been lost with no survivors. Vegas had fallen. Elko and Carson were barely holding. Now, orders came directing the 5th Marines to retake Vegas, throwing the Leathernecks into a battle whose savagery would become legendary in Marine Corps history.

Essential to the outcome were the rapid-firing, accurate, recoilless rifles. It was a killing job, however, packing the 75mm shells over the Korean landscape of hills and paddies to their firing positions. This the Marines had learned during their earlier battle for Un-Gok Hill. At Un-Gok, troops packing ammunition for the recoilless weapons had been barely able to keep the thundering guns supplied in the intense exchange.

After experiencing Un-Gok, Lieutenant Eric Pederson, who'd been a horseman since his boyhood days in Arizona, had asked for, and received, permission to buy and train a horse to pack the ammunition. On a trip to Seoul, he had made the rounds of the war-impoverished stables at the city's racetrack. He had known what he wanted – a strong, swift animal as rugged as the native terrain.

Pederson's keen eye had bypassed most of the lean, hungry horses offered him. He had been about to give up; Then he spotted the dainty-footed, red Mongolian mare. Her name was Ah-Chim-Hai, which in Korean meant "Flame of the Morning." She sported three white stockings and a blaze, and at five years old stood a small 14 hands high. She had been tied along a mud wall, and when Pederson had stretched out his hand, she had pulled to the edge of her rope in greeting.

Flame of the Morning held promise of early speed in her compact body, and her owner had been reluctant to sell her. Pederson had known he'd found his horse, though, and – $250 later – the mare had made the trip back to the base with him.

The weapons unit had renamed her Reckless, and the spirited young mare had soon charmed the Marines. She ate the food the troops ate (including bacon and eggs), invited herself into their tents when it rained, and followed them into bunkers. Marine Sergeant Joe Latham had taken over her training and she had learned not to shy under fire, becoming accustomed to the crashing back-blast of the recoilless rifles. She had been taught to take cover from enemy fire, negotiate communication lines and barbed wire, and carry the heavy ammunition packs.

During her off hours, she had gained a reputation for downing white bread, beer and an occasional poker chip.

On the battlefield, the staunch horse had proven herself a true warrior. In an anti-tank company, the recoilless rifles were moved from unit to unit whenever their firepower was needed. From the first, Reckless worked with the company when they were rushed to the front lines. Hearing the noise of incoming artillery or mortar fire, she was known to bolt for the nearest bunker when off duty. On duty, however, Reckless never ducked when the pack was loaded on her back.

In the battle to retake the Vegas outpost, the thinned Marine ranks had difficulty keeping sufficient ammunition moved up for the constant firing of their weapons. Reckless had been called into action and led over the supply route to the roaring recoilless positions by Sergeant Latham. Without her, there would have been little hope of keeping the rifles supplied.

Wounded now, Reckless panted, arched her neck and kicked out with her front feet, straining to rise up the steep, rocky slope of the ridge. Saddle leather groaned. Another mortar shell slammed to earth, impacting close to her left. Artillery fire cut into a nearby crest.

Nearing her destination, the mare rounded an outcropping and slowed at the approach of a column of Marines. Several were bandaged and limping, supporting one another along their cautious descent. Two of them carried a stretcher between them, where their bloody comrade lay. Reckless knew them. These were the men she'd eaten with, slept beside and nuzzled affectionately with her soft muzzle when wanting attention. She looked in their direction and blew softly, but didn't pause.

Reckless

The trail got steeper. The sound of the battle roared, and at last Reckless pulled up under a small ledge, breathing hard. The final incline to the rifle positions rose at a preposterous 45-degree angle.

Surrounded by the hell of battle, Reckless lumbered toward the rifles and was greeted by a sweaty, streak-faced gunner. Grateful Marines rapidly unstrapped her precious burden. This was her fifth trip to the forward positions, but her first without order or direction. She'd made it entirely on her own.

The weight was lifted from her, and Reckless turned to clear the gun emplacement and head back down the hill for another trip to the munitions dump. But a hand reached out to stay her. The Marines had noticed her wounds.

Shouts went up for a medical corpsman. The medic dodged toward her as a fresh series of artillery blasts scorched over the ridge. Without a second thought, the nearest Marines pulled off their flack jackets, flinging them over the exposed horse. Together horse and men waited out the intense barrage. The corpsman applied a cold compress to the oozing gash on her flank, but painful as it was there was nothing he could do for the nick on her eye.

At last the barrage lifted. Reckless was released for the descent, and with her went the hopes of every man there. Each knew her courageous trips were saving lives.

On that one day, the brave, devoted horse made 51 solitary runs between the munitions dump and the front lines, each time lugging more than a third of her body weight across hostile ground being swept with intense shellfire. She was so effective that she kept three recoilless rifles in action; one able to fire so fast that its barrel overheated and the weapon had to be replaced.

The battle raged for three days, and with Reckless' help, Vegas was retaken and held. She'd covered a total distance of 35 miles on the first day alone, packing more than 9,000 pounds of explosives to the men she'd come to love.

The Marines did not forget her gallantry. After the battle of Vegas, her actions were reported in the press back in the States and were heralded in *Life Magazine*. She was offered a spot on *The Ed Sullivan Show,* and became the heroine of the book, *Reckless, Pride of the Marines.* The Marine Corps brought the gallant horse home to serve out her days at Camp Pendleton, California, where a monument was later erected in her honor.

Finally, for her unmatched loyalty and complete disregard for personal safety, leading to the saving of countless lives, the little red pony was promoted – with full ceremony – to the rank of Marine Staff Sergeant. On hand to receive the honor with his mother was the mare's new colt, Private First Class Fearless.

Peggy

Peggy, an 11-year-old fox terrier, was sleeping in the back room of the Hickory Street store owned by her master, Joseph Coltraro. The little white dog, whose black face fit her like a mask, had lived in New Orleans, Louisiana, all her life and was accustomed to the myriad of sounds and smells around her. The strange hissing and crackling which woke her in the dead of this December night in 1931, however, was unfamiliar.

Perking her small ears forward, she stood and trotted across the wood floor toward the source of the strange noises. Peering into the main store, she was met by an ominous wave of heat and flames. The building, which housed both the store and the family's apartment above, was on fire.

Barking sharply, Peggy ran through the back room, scuttling up the steps to the side entrance of the Coltraros' sleeping quarters. Clearing the stairs, the aging dog flung herself through the apartment to the Coltraros' bed and began pawing earnestly at their covers.

Ordinarily, Coltraro would have scolded Peggy and gone back to sleep, but just a few weeks before he'd seen his dog do an astonishing thing. Peggy had been nearby when her pup was hit by a car. The pup was dying and more cars were bearing down on the spot where she lay. Without hesitation Peggy raced into the roadway and dragged her pup to the curb.

Coltraro, recognizing the uncharacteristic urgency Peggy had shown that day, came instantly awake. He flung back the covers and pulled on his robe.

Following the terrier back down the stairs, Coltraro found fast-spreading flames threatening to engulf his store. Alerted in time, he was able to summon help and extinguish an inferno that would not only have destroyed his livelihood, but perhaps his life and the lives of his family as well.

The bandit-faced fox terrier's heroism did not end there. One day several months later, Peggy followed four-year-old Perry Coltraro when he wandered into the rural area near their New Orleans home. The young boy headed down a gully and straight to a small stream.

On the stream's bank he paused and poked a stick into the mud, watching the water swirl around it. Abruptly he tossed the stick away and stepped back into a clump of brush. There was a loud metallic slam as the jaws of a hidden raccoon trap slammed around his leg.

Perry screamed in agony, thrashing uselessly in the trap's steel teeth, his movements only worsening the terrible wound. Blood began oozing into the dirt where he fell, sobbing, his cheeks pale with shock.

Peggy rushed to Perry and worriedly licked both his cheeks. She wriggled up next to the writhing boy and nuzzled him gently. Perry calmed with her presence and finally lay still. Peggy sat up, but when she did, Perry wailed again, trying to pull the heavy trap apart. Peggy moved close a second time and the boy fell back, exhausted, dirt and tears streaking his face.

Peggy's rough tongue shot out again and again to comfort him. When she again sat up and began to leave, this time the boy seemed to understand. Near Perry's pinned leg the growing pool of blood was beginning to spread over the ground. Peggy would have to bring help quickly.

Slowly Peggy stood up. With a final scrubbing of the little boy's face, she edged away, then turned and raced from the creek. Flying up the incline, she headed toward the Coltraro store in town.

On Hickory Street, Peggy dodged traffic and pedestrians in a whirlwind race against time. Reaching the store, she hurtled through the open door and set up a loud, persistent barking. Coltraro came from behind the counter, frowning. By now he understood. Peggy knew something was wrong.

Frantic, Peggy ran ahead of her master, guiding him back to where Perry lay bleeding to death. She led him down Hickory Street and beyond it to the woods. Not knowing what he'd find, the man followed the small dog with growing apprehension. Stumbling through the gully after her, he saw Peggy halted in front of a small, bent form. Coltraro gave a cry and sank to his knees by his son. Trembling but still, Perry lay waiting for help.

Thanks to Peggy's quick warning, Coltraro was able to spring the trap and rush Perry to the hospital, where the boy recovered from his wounds. For the jaunty, mask-faced fox terrier's courage in saving her family from fire and in leading Coltraro to his trapped son, Peggy was awarded the Latham Foundation Gold Medal Award.

Chester

Deepening folds of skin under the soft fur of his heavily jowled face gave the Chesapeake Bay retriever a worried look, even when he wasn't. Bottomless brown eyes added to this grave countenance. But one day in 1978, as Chester – nicknamed "Chessie" by his owners – watched five-year-old Kenny Homme wander off from the family's backyard in Livingston, Montana, his anxious gaze expressed real concern.

Kenny's mother, Mrs. Homme, had been checking on the boy periodically through the kitchen window while she went about her work. Bent over a sink of dishes, she hadn't lifted her head in time to see her son leave the yard. Beneath the turquoise-blue Montana sky, the adventuresome boy walked curiously toward the brink of a steep hill behind the home. Dutifully, Chester followed.

The incline ended in a fast-moving creek, dangerously swollen with springtime runoff. Powerful currents grabbed the leaves and sticks young Kenny threw into the torrent, carrying them swiftly away. Fascinated, the boy watched this phenomenon while marching precariously back and forth along the top of the bluff.

Suddenly Kenny lost his foothold in the loose, damp earth. Chester watched as he toppled down the hill, grabbing at handfuls of soil in an effort to stop himself.

"Help! Save me!" Kenny screamed. His frantic face turned once toward the retriever before he disappeared into the creek.

Chester was down the incline in an instant, leaping into the cold water at the spot where he'd seen Kenny go in. Already the swift current had taken the boy out of his reach. Chester saw Kenny surface downstream, still shouting for help.

Mrs. Homme heard her son's cries. Dashing quickly from the house at the top of the incline, she was met with a horrifying scene. Chester was battling desperately against the current, while Kenny was being swept with grim certainty toward a metal culvert which siphoned water into another channel.

Chester didn't make it to Kenny in time. Caught in a strong undertow, the boy was dragged into the tunnel, the echo of the dashing water drowning out his cries for help. The dog, swept past the culvert's opening, lunged for the spot where Kenny was pinned, but now the water worked against him. Again and again he swam toward the boy, only to be swept hastily backward.

Chester and Kenny had been in the water for five minutes. The surging waves slapped at their faces, choking them, and the tearing force of the current was relentless. As the precious seconds ticked by, the dog fought to reach the struggling boy, who was fast losing his remaining strength.

Once the courageous Chesapeake got so close to Kenny that the small boy reached out and grabbed his fur. But as soon as the dog kicked toward shore, Kenny lost his grip. Immediately Chester turned and headed for the culvert again, and again Kenny's near-frozen fingers sank into the wet fur. Once more the raging water tore them apart.

Chester snorted water from his nose, his breath coming in gasps, and turned back toward Kenny a third time. Heaving forward, he swam beneath the boy's small struggling body. This time, with the dog holding as still as he could in the powerful current, Kenny was able to slip his tired arms around Chester's neck and climb onto his back.

Almost exhausted, Chester eased out of the culvert with his precious burden. He swam downstream, maneuvering with the current until he finally reached the shore. Mrs. Homme pulled her son into the shelter of her arms as the devoted retriever splashed from the freezing water.

"If we didn't have Chessie, we wouldn't have a son." Mrs. Homme later remarked, describing the rescue. For his vigilance and courage in saving little Kenny's life, the Chesapeake Bay retriever received the coveted Ken-L Ration Dog Hero of the Year Award for 1978.

Bill

It was a clear afternoon in 1926 as the big ore train rumbled toward the blind curve that marked the approach to the Washington mining town of Lucerne. Yanking the signal whistle cord, the engineer gazed out across the countryside at small, shanty-like homes with their littered yards and sagging clotheslines.

Around the bend, Bill, a lanky, patch-coated English pointer, was alerted by the whistle and perked up his long ears. The train was still invisible behind the mountain, but in minutes it would come clamoring directly behind the home of his owners, the Moore family.

Mrs. Moore, out hanging clothes, looked around to check on her two-year-old son, Richard, who'd been toddling about the unfenced yard. Accustomed to the constant passing of trains, and satisfied that Richard was at a safe distance, she had kept her back turned a moment too long. She hadn't seen Richard ambling toward the tracks.

As the sound of the approaching train grew louder, she searched the yard with growing apprehension – but Richard wasn't there! She called his name, only to have the approaching train's roar drown out out her words.

Suddenly she spotted him, oblivious to danger, playing on the tracks between the humming rails.

The train was so loud now the ground shook. The air reverberated with its steaming breath. Her little boy seemed impossibly far away; how could she reach him in time?

At almost the same instant, Bill saw the boy. As Mrs. Moore began a desperate run, the pointer streaked across the yard toward the tracks. The earth flew beneath his feet. Richard, his small round face turned down the distant tracks, was totally oblivious of the onrushing engine.

The train roared forward. Stumbling, Mrs. Moore caught a glimpse of the ore cars, seeing flashes of daylight between them as they rumbled by. She realized, with horror, that she could not reach the boy in time. It was up to Bill.

Locked in a race with death, Bill and the train both rushed with terrible certainty toward the spot where Richard stood. Helpless to stop the train in time, the engineer frantically yanked the whistle cord. Richard jumped and began to cry. With his back toward the huge train, he raised his arms in confusion, believing his mother close enough to snatch him to safety. The shattering vibration of the railbed shot up through his feet as the barreling, blind eye of the locomotive bore down on him.

Suddenly there was a flash of brown and white. Bill, streaking over the ties, missed the onrushing locomotive by inches and slammed into Richard, knocking him into a nearby ditch.

While the train pounded past beside them, Bill stood panting heavily, his front paws firmly planted on the boy's back, refusing to let him move until the big rumbling cars had finally passed.

Mrs. Moore, frantic and crying, reached them at last. Gathering Richard up and hugging Bill, she knew without doubt that the dog's quick action had saved her son's life.

The big pointer's loyalty, intelligence and courage did not go unnoticed. The story was carried in the newspapers, and Bill was nominated as a recipient for the Latham Foundation Gold Medal Award.

Bill

Kitty

Eva Chesney, Kitty's 70-year-old mistress, walked with a cane. For the few short months of Kitty's young life, the mixed-breed cat had lived with Eva and had taken to dodging in and out of the elderly woman's walking stick as Eva made her resolute way around her Stevens Point, Wisconsin, home.

Late that Christmas Eve in 1986, Kitty was underfoot as usual, running circles around Eva's feet and dashing amiably before her as Eva moved to feed small logs into her wood-burning furnace. "Merry Christmas," Eva remarked to her sole companion, pausing to give her four-footed friend a celebratory pat. Then, with the fire well stoked, she headed up the narrow stairs to bed.

Kitty followed her elderly mistress until Eva reached the landing and turned toward her bedroom door. The kitten would have gotten into bed with Eva, but the woman scolded her lightly. Who could sleep with a playful cat pouncing over the covers? Eva shut her door gently but firmly, and Kitty began her nightly prowl of the aging, turn-of-the-century house.

It was a cold night. Outside, snow lay in small drifts under an overcast sky. Eventually tiring of her inquisitive wanderings, Kitty curled up on the rug, lulled by the warmth from the now-blazing furnace.

When she came awake at about 4:30 a.m., she was already feeling the first lethal effects of smoke inhalation. Blazing away, the furnace had overheated and smoke was pouring from the chimney into a nearby basement laundry chute. From here it spread greedily into the downstairs rooms. It was a silent suffocator, more deadly than fire itself, and it was beginning to fill the stairwell leading toward Eva's bedroom.

Kitty struggled upright, gripped in the deadly fumes' intoxicating spell. The young cat stood and shook her head, sneezing and blinking her eyes to restore her normally acute vision. Dense smoke filled the rooms, collecting in heavy clouds along the ceiling and in passageways leading to the house's second story.

Still coughing and sneezing, Kitty crossed the room to the stairs leading to the rooms above, where Eva still slept. In her weakened condition, the steps the cat had bounded down easily only hours before now loomed huge and forbidding. Slowly she put her paws on the first riser and, with smoke-filled lungs heaving, began an ascent which would be snail-paced and painful.

Finally she was outside Eva's bedroom door. There wasn't much time. After the difficult climb she could barely stand upright. Pacing and meowing weakly, she listened through the wood for any sound of Eva rousing. There was none. Then, in desperation, she faced the blank door and began clawing at the wood.

Soon there was a noise from within. Alerted by Kitty's frantic meowing and scratching, Eva arose from the bed and searched for her cane. Kitty heard the familiar thump-thump on the floorboards. Then there was a rush of air as Eva opened her door and gasped, choking on the fumes. Kitty yowled anxiously as she stumbled from the room.

Now they had to get down the dark stairway and Kitty, who had been in the deadly smoke for some time already, was so weak she was barely able to crawl. As Eva, who was also becoming weakened by the overpowering fumes, fumbled down the steps, Kitty tracked her by following the sluggish thumping of her cane.

With leaden limbs, Kitty made it to the last step — and that was all. Impenetrable smoke swirled black around her as the exhausted little cat collapsed on the downstairs floor. Eva kept thumping on toward safety, assured now of reaching fresh, life-giving air.

When Eva Chesney lurched alone gasping into the sharp, snowy dawn, there was no doubt in her mind that her cat's signal had saved her life. The next day, the story of Kitty's heroic rescue was circulated nationally in newspapers across the country, and the young, mixed-breed feline was cited for giving the greatest Christmas gift of all – the gift of her own life to save someone she loved.

Blaze

When the sun heated his fur coat, the small, cream-faced collie retreated under the porch of the Hecox's farmhouse near Timewell, Illinois. As he lay panting in the shade that day in 1957, he saw a movement out of the corner of his eye and recognized two-and-a-half-year-old Dawn Hecox toddling across the yard. With a sigh, he lowered his head on his forepaws.

Wandering away from her mother, Dawn had caught sight of the new litter of shoats rooting in the farm's pigpen. Fascinated, she slowly made her way toward them. As she crawled through the hole beneath the bottom board, the mother hog spied her.

Known for her ornery nature, the big sow had become extremely protective of her litter. Now she grunted her displeasure over the little intruder, her eyes narrowing as she calculated the distance between them.

Dawn, oblivious to the threat, toddled laughingly toward a lone shoat. The sow, infuriated, lowered her head and began her charge. Dawn felt, too late, the trembling of the earth as the huge animal barreled down on her, and turned just in time to take the brunt of the gigantic snout.

Ramming her full weight into the little girl, the sow knocked Dawn against a fence post. The little girl's head snapped backwards and her mouth spewed blood. As Dawn crumpled helplessly to the ground, the sow's yellow teeth ripped at her. Sharp, mud-caked hooves sliced into the child's tiny body.

Under the porch, Blaze heard the sow's high-pitched screams of rage and leaped to his feet. He'd distrusted and feared the irritable beast since puppyhood, when a run-in with her sharp hooves had sent him spinning. From then on he'd given her a wide berth whenever their paths had crossed. Now, however, a glance in the direction of the pigpen brought the sight of little Dawn being mauled and trampled.

Scrambling out from under the porch, Blaze dashed across the barnyard, scattering chickens in all directions. With a snarl he cleared the fence and flew at the sow. Lunging for her throat, he sank his teeth into the thick, bristly skin, and hung on.

The sow squealed angrily and swung Blaze sideways, slamming him into the fence. The impact loosened his hold and he dropped to the ground, stunned. Just as the furious pig lunged at Dawn again, Blaze shook himself and leaped forward.

Coming inches from the hated hooves, the collie clamped his teeth deep into the hog's flank. Enraged, the sow now turned her full fury on the dog.

Dust billowed and the frightened shoats ran squealing to the corners of the pen as the snarling, bellowing tangle of dog and pig whirled around the enclosure.

Blaze was no real match for the big hog, but his steady onslaught at last bought Dawn enough time to roll under the fence to safety.

Mr. and Mrs. Hecox, alerted by the clamor in the pigpen, instantly realized the absence of the toddler and rushed to the site of the battle. Spotting the torn and bleeding Dawn, they ran to their daughter's side and Mrs. Hecox gently picked her up. Mr. Hecox jumped into the pen to separate the still-warring animals.

Dawn was sped to the emergency room of a nearby hospital, where it was found she'd been severely bitten and four of her teeth were missing. For two days she remained in critical condition, and she required constant care for three weeks before fully recovering from her ordeal. If not for Blaze's fearless interference, she would surely have died from the sow's mauling.

For his life saving act, Blaze became the second collie ever to win the Ken-L Ration Dog Hero Award.

Cher Ami

The Battle of The Argonne, fought in September/October, 1918, was one of the most bitter of the war. Cher Ami, a small, slate-colored racing pigeon, became a legend during that epic struggle. Donated by Great Britain and trained by the American Signal Corps to carry messages over the battlefield, like many other specially trained birds of his kind, he was often the sole means of communication between troops during battle.

Locked in combat with the Germans, the U.S. 77th Division spearheaded the American attack. On October 2, the seventh day of fighting, Cher Ami and four other pigeons huddling in their cage were selected to accompany the 308th Regiment into combat. This unit of 700 men, commanded by Major Charles W. Whittlesey, advanced more quickly than the other troops and soon penetrated deep into the enemy's front.

The shock of explosions and roar of the guns surrounded and pounded the crouching birds. Whittlesey's men were holed up in a pocket of forest too dense for accurate artillery support. But when the Major relayed his position, he was ordered to advance again and, in a series of perilous moves, led his men up Hill 198 and down the other side.

In their determined attack, Whittlesey and his battalion had advanced beyond their flanking forces of the 307th. The Germans had them surrounded and were smothering them with mortar and machine gun fire. Suddenly the last message they would receive from headquarters came through: "Do not advance." But it was hours too late.

At dawn on October 3, the Germans renewed the attack on the besieged Americans. Cher Ami was still huddled in the basket with the other pigeons. The battalion was dug in, fighting back heroically, but being blasted by German "potato masher" hand grenades. Sergeant Omar Richards, Cher Ami's handler, released a first and then a second pigeon, both with messages requesting artillery support, ammunition and medical supplies. The birds never arrived at headquarters.

Casualties were heavy by the dawn of October 4. Food was running short and the only water source was covered by German snipers. A party who'd tried to break out and go for supplies had been stopped, and one runner captured. A third pigeon was sent aloft with the news.

The battalion, though taking a far greater toll on the enemy, had dwindled to 245 men. The captured runner was returned with a note from the German commander demanding surrender. The demand was immediately rejected! Then, in the worst hours of the siege, American artillery suddenly began shelling their own sector.

Shell-shocked and exhausted, Richards inched to the pigeon basket and withdrew the next to the last bird. Very likely the first three had been picked off by German marksmen. The 308th's only hope was that one of the two remaining birds would get through.

But Richards fumbled with the precious bird, and before he could insert the message in its leg tube, it flew free. Richards then reached for the last pigeon in the basket, Cher Ami.

The message was slipped into the small metal tube attached to Cher Ami's right leg and the tube was capped. It gave their position and the plea: "For God's sake, lift your fire!"

Set free, Cher Ami flew above the battlefield, but after a few quick spirals, he settled in a nearby tree and began to preen. Only when Sergeant Richards risked his life by dashing through the barrage of machine gun fire to flush him out did Cher Ami move on.

Airborne again, the little pigeon circled to pick up direction. It was over 20 miles through enemy fire to home. Shrapnel and chunks of bursting shells flowered their murderous patterns in the air around him. The bullets of German riflemen and machinegunners cracked and snapped.

Suddenly a bullet crashed against his head, tearing out his left eye. Stunned, the pigeon tumbled toward earth. He was only semiconscious when shrapnel tore into his chest, destroying his breastbone.

He plummeted, the nightmare worsening as the earth reeled toward him. But the compass in his head, the instinct for home, fought back. He regained himself, and flew on.

Death surrounded him in the sky. Cher Ami was struck a third time, his lower right leg ripped away by more flaming shrapnel. When the brave pigeon finally limped into the safety of the loft, he was a mass of blood and feathers. But he'd made it! Miraculously the message was still there, hanging from the torn ligaments of his remaining leg.

Immediately the order went out to the American artillery batteries firing on Major Whittlesey's position, and the lethal fire ceased.

When Cher Ami recovered from his wounds, he was shipped to the United States aboard a naval warship in a "pigeon officer" cabin provided for him by General Pershing. He lived the rest of his life as an honored pet of the Signal Corps in Washington, D.C.

As for Major Whittlesey and his men, they fought their way out and Major Whittlesey was awarded the Congressional Medal of Honor. For his part in helping save the "Lost Battalion," the heroic pigeon was awarded the Croix de Guerre, the French War Cross, for bravery.

Cher Ami

Taffy

The barn and stables on the shores of Idaho's Fernan Lake stood placid under a cold, clear sky that day in 1955, with nothing threatening in the air. Taffy, a small, honey-colored cocker spaniel, played with her three-year-old pal, Stevie Wilson, while Stevie's father, Ken, worked a saddle horse nearby.

To ensure that the pair kept out of mischief, the elder Wilson, before taking his mount into an adjoining corral for a workout, had secured the boy and dog in one of the unused corrals and carefully latched the gate. The two were content until an unsuspecting neighbor came along and released the latch, giving young Stevie his chance to wander.

Once the gate swung open, Taffy and Stevie, both short-legged, trotted together into the bright afternoon. Without much decision, Stevie headed straight for the icy waters of the lake, which lay enticingly smooth-surfaced and blue just down the slope from the corrals. As they negotiated the slope, the sounds of Ken working the horse grew distant.

The lake's water fascinated Stevie, who toddled closer and closer to its shore. Taffy, who mistrusted it, hung back while the toddler edged along the small waves, first getting his feet wet and finally wading in. Watching, the cocker nervously paced back and forth.

Suddenly Stevie stepped in a hole and went under. The dog halted her nervous walk and barked excitedly, but the little boy didn't surface. Taffy ran halfway up the incline leading to the corral, then back to the spot where she'd kept vigilance on the shore. Still Stevie didn't appear. Slowly, his red mackinaw floated to the surface.

Taffy barked again, darting toward the water. No Stevie. Quickly she sped up the incline and into the corral where Ken was working the horse. Yipping as loud as she could, she advanced on the pair.

The horse's flanks were sweaty, its nostrils flared. A high warning whinny pierced the air. Ken, sure that Stevie was still safe in the enclosure, shouted to Taffy to get back. One stab of a steel-shod hoof could kill a dog Taffy's size, or get Ken himself thrown.

Frantic at not being able to communicate, Taffy ran out of the corral and back down to the lake. She plunged into the frigid water and swam toward Stevie. The cold had her small body shaking almost immediately, and she trembled as she circled the boy. Desperately, she tried to get her teeth into the red mackinaw, but she couldn't move the seemingly lifeless Stevie.

The attempt was futile and, tiring, Taffy abandoned it and swam hurriedly to shore. Out of the water, she again skittered up the incline to the corral.

63

Scooting under the fence, the little cocker saw the saddle horse looming high above her, the hard ground vibrating from the drumming of its heavy hooves as she approached. Narrowing her eyes, Taffy scrambled through a swirl of choking dust to pursue Ken's mount. She barked incessantly, dancing between the confusion of legs, the massive hooves so near they brushed her ears.

Ken shouted at her, scolding, but Taffy kept up the harangue as she nipped at the tender skin above the horse's hooves. She moved in for a second nip, barely slipping out of reach as the angered horse kicked out.

Ken, angry too, concentrated on the horse, trying desperately to keep his seat in the saddle as Taffy, with even more desperation, tried to unseat him. When Taffy moved in for the third time, the horse reared and screamed, aiming a kick through the dust directly at the dog. Taffy, exhausted by her efforts, heard the keen whistle of the steel-shod hooves slicing the air as she slipped out of reach just in time.

Still she stubbornly stood her ground, barking loudly, and this time when Ken looked down, he took in her soaked coat and anxious eyes with sudden understanding.

A shout to the neighbor confirmed that the gate had been opened. Ken swung down from the saddle and ran behind Taffy to the lake.

Stevie's father saw the red coat almost at once. Jumping into Fernan Lake, he pulled his unconscious son from the bottom and began artificial respiration. The alerted neighbor called an ambulance, and Taffy waited.

It was six hours before Stevie Wilson regained consciousness. A few more moments in the freezing water and he wouldn't have come out alive. Taffy's swift action and bravery in facing down the horse had saved the little boy's life.

In honor of her heroic determination, the devoted cocker spaniel was awarded the Ken-L Ration Dog Hero of the Year Award.

Dick

Dick, a burly German shepherd, was moving toward his master's gold mine and arrived just moments after the brush fire which had begun nearby in the remote woods outside Downieville, California. Fanned by hot winds, the fire leap-frogged through the dry forest of the famed Mother Lode region at an alarming rate.

Inside the family-run mine were Dick's master, Herman Stark, and Stark's elderly father, both working well down a 1,000-foot, air-vented shaft. With temperatures reaching the 90s that August afternoon in 1931, conditions were extremely dangerous.

In minutes the well-fueled fire swept up to the timbers at the mine entrance. As the old timbers ignited, spreading flames across the miners' only avenue of escape, volumes of smoke began pouring down the shaft and threatened to asphyxiate Stark and his father.

Alarmed, Dick paced in front of the burning mine entrance and barked loudly. A close companion to Herman Stark, he was no stranger to safeguarding his master. The year before, on the way home from working the mine, Stark had been drawn to a small bird clutching a tree limb and gripped in a kind of hypnotic trance. Not realizing the source of the bird's stupor, Stark had approached it out of curiosity. Only when Dick stepped protectively in front of him did he see the snake.

A long timber rattlesnake, already poised to strike, lashed out, and the German shepherd took the venom meant for Stark. It had been many days before the big dog fully recovered from the near-fatal bite.

Now Dick tried to warn Stark again, but his anguished howls went unheard. Laboring to the din of picks and shovels, the two men were unable to perceive any sounds coming through the mine's long tunnel.

The heat and flames were growing intense, swirling the air at the entrance to the shaft. Sparks shot skyward as the timbers blazed like torches. Soon the fire formed a wall of flame across the space where Stark and his father would have to come through. Dick stopped barking and leaped through the flames. His fur was instantly singed black.

Landing short of his mark, Dick pitched into the edge of the fire, burning his paws as he scrambled across it. His momentum carried him into the hard rock mine's granite wall. Picking himself up, he watched a heavy roll of smoke flow ahead of him toward the shaft. Dick ran with it deep into the bowels of the earth.

His paws throbbed with pain. Sharp rocks and clods of earth reached up to slice the now-tender pads. He rounded a corner where the poisonous fumes were so heady he stumbled and nearly blacked out. The shadowy mine shaft was as still as a tomb.

Dick slowed, confused. The smoke was distorting his sense of smell. He cocked his ears for the familiar sounds of the Starks.

Dizzy from the intoxicating smoke, Dick headed further down the shaft. As he reached the lower levels, the air was less thick and he regained some of his former agility. Rounding another corner he heard Stark's pick ringing against stone and, trembling, he entered the small cave where the miners were working.

Dick's appearance said it all. The Starks took one look at his blackened, smoking coat, his torn and blistering paws, and prepared for their ascent up the shaft. Taking their shirts off and throwing them turban-like around their heads, they followed Dick.

Once clear of the air pocket at the end of the tunnel, every step was a nightmare. The smoke poured down to meet them in black, suffocating sheets, heat from the fire advancing with it. Haltingly, they began the last of the 1,000-foot climb, with Dick in the lead.

Suddenly the elder Stark stumbled against a wall, hacking from the smoke in his lungs. Nearly lost in the haze, Dick turned back to him and waited until the aging man could stand upright again and continue the wearying climb.

As they neared the entrance, both men staggered back from the furious fire. Burning timbers sagged and snapped around them. Dodging flaming chunks of wood, Stark called to his father, "We'll have to run for it!" Then, arms up to shield their faces, the two men dashed into the flames.

Dick paced back and forth inside the flaming entrance and whined anxiously as his master disappeared. At last he heard Stark calling to him from the other side. As he leaped toward his master's voice, a loud crack came from behind, followed by the thunder of falling timbers as the entryway collapsed. A few seconds more and none of them would have left the mine alive. Dick's devotion was formally recognized by the Latham Foundation Award for animal heroism.

Woodie

Panting a little from the steep climb, Woodie sat attentively beside her mistress at the top of a forested path in Cleveland's Metroparks Rocky River Reservation. Her mistress, Rae Ann Knitter, 28, threaded one hand through Woodie's collar to keep her from following Rae Ann's fiancé, Ray Thomas, 24, as he scrambled up a steep shale incline.

For Ray, an amateur photographer, the cliff's top promised a spectacular view of the river running near Cedar Point Hill on the reservation, and he had the advantage of a fine autumn clarity that afternoon in September, 1979.

From their place on the path, Woodie and Rae Ann waited, watching the young man gamely scaling the incline, camera in hand. Bits of flat rock skittered and rolled beneath Ray's feet as he neared the top of the bluff. They saw him position himself for the shot, take a step – and disappear from sight.

Woodie had been with Rae Ann for six years, ever since the young woman picked her out of a batch of free pups in Columbia Township and took the fluffy little collie-shepherd home. Grown up, the dog had a reputation for being friendly and, in particular, obedient. Never before had she acted as she was about to now.

Later, Ray would describe his horrifying plunge as beginning almost 15 feet from the edge of the 80-foot cliff, in loose shale – with nothing to grab onto. Halfway through his long fall he lost consciousness. His fiancée waited, unaware of his peril. It was Woodie who knew immediately that something was terribly wrong.

With Rae Ann's hand still firmly on her collar, she began to twist and tug in a sudden frantic need to be free. Her sharp ears had told her the story. Ray was falling over the cliff! Woodie had to get to him!

When Rae Ann relented and released her, the collie-shepherd ran quickly up the same bluff Ray had climbed moments earlier. At the top of the precipice her fears were confirmed. There, 80 feet below, lying face down in shallow water, was Ray's motionless form.

Like tiny bombs, pieces of loose shale tumbled away into the air beneath Woodie's splayed paws. Woodie gathered herself on her haunches, barked once, and leaped into space.

Rocks fell along with her – down, and down. Her legs flailed for a non-existent grip in thin air. The ground slammed up too fast. In shock and pain, Woodie found herself at the bottom of the cliff, several yards away from Ray. She tried to stand, but the impact had broken both her hips. She yelped in pain. Her back legs would not work, yet Ray still lay helpless, the water swirling around his nose and mouth, suffocating him.

Ignoring the agony and trauma of her own wounds, Woodie began dragging herself toward Ray's inert body. Using her forepaws, she clawed over the uneven ground inch by inch.

The minutes seemed like hours, but at last she was at Ray's side. The terrible fall had broken his back and one arm, and Woodie moved carefully, sensing these injuries. Gently she nosed the now semiconscious man. As cautiously as if he had been a newborn pup, she pulled his head from the water. She heard Ray gasp and cough, breathing in precious air. She nudged him again, giving encouragement, then dropped down protectively beside him to wait.

Rae Ann, still on the cliff above the stricken pair, was able to summon help from some nearby hikers. Woodie watched over Ray until an ambulance arrived, and only then did she allow Rae Ann to scoop her up in her arms and tend to her own serious injuries.

There was no doubt that the amazing courage of the long-haired collie-shepherd saved Ray Thomas from drowning in shallow water while he lay paralyzed after his fall. Woodie's heroic dive to his aid, without worry over her own safety, earned her national attention. In October, 1980, she received the Ken-L Ration Dog Hero of the Year Award, accepting it in Los Angeles. She was accompanied by Rae Ann and by Ray, who had recovered enough to walk with the use of canes.

And her touching loyalty prompted a popular television show to award the couple whose love Woodie had risked everything for a special prize: an I.O.U. for one honeymoon in Hawaii.

Woodie

Rex Beach

Rex Beach, the tall, lean saddle horse, had wandered deep into the Washington woods near Port Angeles after leaping a corral fence one night in 1931. Then, having grazed on the tender grass shoots he loved so well, the horse halted in a moonlit clearing and raised his head. He heard approaching footsteps and recognized a man's familiar smell.

Less than a quarter of a mile away, with lead rope in hand and tracking the horse, was Rex Beach's owner, Walter Devereux. Walter was finding his horse's trail relatively easy to spot thanks to a full moon, but the rough terrain made the going slow. Pausing under a rock promontory, he listened for any night sounds filtering past the heavy-limbed trees.

Now a second, less-welcome scent came to Rex Beach on the breeze. It said cougar, an animal all the horse's instincts told him to avoid. He gave a nervous whinny and left the clearing, heading in Walter's direction.

Rex Beach spied the big cat stalking Walter in stealthy silence. Crouched low, muscles taut, the cougar had managed to close the gap between himself and his quarry without a sound. At the last moment before the cat sprang, Rex Beach charged.

Within seconds, horse and cougar were locked in a deadly duel. Ears flat and nostrils flared, Rex Beach bore down on the feline assassin. The big cat, caught off guard, turned and screamed in startled wrath, darting under the raised hooves and lashing out.

Walter whirled around in shocked surprise. Until he heard the wild cat's defiant scream he hadn't realized that, as he'd been tracking his horse, he was also being tracked. Terrified, certain of attack, he looked helplessly around for a weapon. Finally deciding on a few fist-sized stones, he turned to watch an incredible drama unfold.

Infuriated, Rex Beach rose up again and came down, hooves flailing. With lightning speed the cougar withdrew, snarling, but the respite was brief. Again the cat sprang forward, giving the horse barely enough time to whirl and face him. Instantly Rex Beach kicked out, catching the cat with a blow that knocked him backward.

Spitting furiously, the cougar rolled upright and leaped for the horse's jugular. Rex Beach sidestepped, but not in time to escape a vicious swipe of powerful claws. Blood flew. The cat rushed forward, teeth bared, but Rex Beach reached out at the last moment with his own powerful teeth, grabbing the cougar by the neck and shaking him hard.

Flinging the big cat away, the horse pawed the ground. Now his adversary was bleeding too, his snarls coming between gasps for breath. As Rex Beach went on the attack again, stomping, the cougar retreated up a tree.

"Rex Beach!" At the sound of his name, the horse turned to see an incredulous Walter standing next to him, unharmed. The treed cougar took that moment of distraction to jump from his perch and disappear into the night.

Rex Beach started to give chase but Walter stayed him, laying a grateful, restraining hand on his neck. The battle was over. The courageous saddle horse had saved his master's life.

Later, Rex Beach's incredible battle to rescue Walter Devereux from the attacking cougar would bring the horse acclaim. His picture and exploits were printed in newspapers throughout the country, and he was nominated for a Latham Foundation Gold Medal.

Muggie

Muggie heard the sickening screech, followed by a resounding thud, as it echoed through the empty, moonless night. The Belgian sheepdog was roaming the grounds of the residence of R.D. "Bob" Spruling, her master, just before midnight one evening in March, 1930. Spurling's home was located south of Everett, Washington.

The noise had come from the rural stretch of highway that ran nearby, and Muggie went to investigate. The 55-pound dog, with her long neck ruff and plumed tail, blended into the shadows as she padded softly along the shoulder of the road, nose down. Cars skimmed by, their startled drivers honking shrilly at her sudden appearance in the headlights. Then the disappearing taillights would leave her in the shadows again.

Muggie heard the groan and smelled the injured man's blood even before she saw him. He was rolled off the shoulder of the road and lay face down in a narrow ditch. High weeds along the shoulder and the dark night conspired to hide him from passing motorists. The man was unconscious.

Albert Jacobson, an Everett resident, had been struck by a car as he made his way home along the highway that night. Immobilized by massive internal injuries, the elderly man lay still and hardly breathing. Muggie approached him with a soft whine.

The sheepdog bent over Jacobson and put her warm tongue along his cheek. She nudged him with her nose but got no response. Then she leaped up out of the roadside ditch and raced for home.

Minutes later, Bob Spurling woke to the sound of his dog making a hair-raising clamor beneath his window. Barking and howling, the sheepdog was running in agitated circles on the lawn. Spurling rose, perturbed, and tried to quiet her.

Speaking harshly to Muggie through his raised window, Spurling ordered the dog to lie down. Instead, the usually well-behaved Muggie barked louder. Spurling had never seen her so upset.

As soon as the light switched on indoors, Muggie leaped for the back door and scratched relentlessly at the screen. When Spurling appeared in his pants and hastily thrown-on robe, she bounded into the pool of light cast from the kitchen behind him. She was still barking and howling. Quickly she ran to the edge of the shadowed yard and back again. Scratching his head at her strange behavior, he grabbed his flashlight.

When he came down the steps Muggie gave a frenzied yelp and raced for the highway. Making sure Spurling was following by constantly looking over her shoulder, she led him into the night.

Jacobson lay nearly three-quarters of a mile from Spurling's place. Out on the road, Spurling tried to give up on what he thought was a ridiculous wild goose chase, but Muggie ran back to him and tugged on his arm with such desperation that he relented. He swung his light along the murky highway, seeing nothing except Muggie with her nose to the ground and loping purposefully ahead.

Albert Jacobson was lying just where she'd left him. Still unconscious, the badly injured man was rapidly bleeding to death, his limbs stiffening and his lips blue with shock. When Muggie reached him, she laid a paw gently across his inert back and then stretched out protectively alongside him.

"My God!" Bob Spurling breathed, shining the light over the crumpled form. He knelt next to the injured man, lifting his wrist to search for the faint pulse. Beside him Muggie's plumed tail beat hopefully in the dirt.

Given emergency medical care and many days of bed rest, Jacobson recovered from the hit-and-run accident and would often visit the dog who had saved his life on the dark, lonely highway. The two had never met before that fateful night, but they soon became good friends. In fact, the Belgian sheepdog fast gained the admiration of the entire neighborhood.

Tragically, Muggie herself was hit several months later on the same highway. She was struck by another hit-and-run driver who left her lying gravely injured in a ditch near the one where Jacobson had lain. She died before help arrived.

Melissa

Melissa had been raised accustomed to the ring of the cash register, the shuffle of feet and the exchange of conversation with customers frequenting the Shadow Record and Book Exchange in Denver, Colorado. These rhythms were so much a part of the three-year-old cat's life that, when the original owners had sold the business three months before, Melissa had stayed on in her usual position of store mascot.

One day in July, 1977, the big-footed, sulky-eyed feline dozed in her favorite corner while Diane Wakabayashi, one of the store's new owners, handled sales behind the register. Melissa had become accustomed to Diane and her partner in the months since they had taken over the shop, and a genuine affection had developed between them. Now, in the regular rush and lull of afternoon business, she could hear Diane helping customers who, one by one, made their purchases and left.

Melissa stretched, now fully awake, and shook her dense white fur. From beyond her resting place came the sounds of a single browser, a stranger, alone with Diane.

Suddenly the customer's voice barked through the store with a menacing snap, and Diane's answer, devoid of its usual friendliness, was frightened and strained. Melissa narrowed her eyes as if the tension in the air were a haze she had to see through.

Leaping soundlessly to the floor, the alerted cat stalked toward the troubled voices. What she saw when she rounded the corner of the aisle made her arch her back in fury. Diane was backed against a wall, and a man was threatening her with a large knife. The man demanded that she open the cash register, enforcing his words by slicing the blade through the air between them.

"Do it yourself," Diane protested, afraid of being nearer the robber and his weapon. Melissa waited no longer.

Tail erect, fur on end and claws flexing in her paws, she sprang to the top of the counter and stalked toward the intruder, punctuating her slow, deliberate steps with a low, throaty snarl.

"Get back!" the startled thief shouted, turning the knife toward her. Its sharp edge gleamed in the overhead light. "Get back!" he repeated, his voice rough with warning. Melissa's eyes were narrowed slits of menace. She took another step... and another. "Get back!" the thief commanded again. Melissa's snarl reached a higher pitch.

The thief flinched, brandishing the knife at her menacingly, but still she came on, closing the gap between them. Melissa advanced to within a few feet of him, her snarl now steady and her eyes locked on his in a glassy stare.

The now-frightened man forgot he held Diane prisoner behind the counter and began looking around for some line of defense against Melissa's seemingly imminent attack. Picking up a nearby chair, he hurled it at the cat, but she ducked under it and advanced again. The assailant, totally distracted, stumbled further away.

At that moment Diane seized the opportunity Melissa had given her and dashed from the store, crying for help. The intruder fled in the opposite direction, so unnerved by the cat's determined defense that he didn't even stop to empty the cash register on his way out.

Melissa's story was carried in the *Denver Post* the next day. Eventually her heroism in facing down the armed robber that July afternoon led to her being awarded the Carnation Company's national Cat Hero Award for 1978.

Melissa

King

I

The big dog woke to the smell of smoke — smoke acrid to the sensitive tissues of his nose and lungs. Through the sliding door left open for him in the house's recreation room, he could see familiar stars poking between the heavy-limbed trees and feel the cold night air. Nothing different there. But from under the closed plywood door that led to the rest of the house, in the bedrooms where his family slept, poured an ominous dark stream.

Realization sparked every nerve of his body, and the German shepherd/ Husky mix named King leaped to his feet. It was Christmas night, 1981, five years to the day since he'd been found by his owners, the Carlsons, a bloody heap — the victim of a gunshot wound — on the doorstep of their Granite Falls, Washington, home.

That Christmas the Carlsons had just lost their family dog to a heart attack, and had felt they'd never be able to replace the beloved pet. They hadn't counted on the injured and abandoned King needing them. King himself had filled his heart since that time with the Carlsons, Fern and Howard, together with their two children, Pearl and Howie.

Howie was sleeping at a friend's that night, but Pearl and Howard, who suffered from a chronic lung condition, and Fern, three of the people he loved most in the world, were closeted somewhere in that lethal smoke. Already he could feel the heat reaching toward him. Yet still from the open sliding door the sweet night air poured in, beckoning. How easy it would be to slip out into the shadows.

King hesitated only a moment before turning his back on the open door and freedom. Instantly he began tearing at the one which barred him from his endangered family.

The hollow-cored door splintered under the attack of his teeth and claws, but did not give way. Angrily King continued ripping at the veneer, gnawing it off in great strips. Spittle flew. The door was hot to his paws and tongue. His jaws gnashed at the wood, embedding several splinters deep in his mouth, but the hole was widening.

By the time he'd chewed and clawed a small opening through the inside layer of plywood, the fire had found it. Sheets of smoke and flame roared toward the gap he'd made. The room on the other side was engulfed in flames and there was no other way to get to the Carlsons but to go through that room. Again he could feel the relief of the night air from the open door at his back.

Ignoring this route of escape, King returned to the plywood. Fiercely he gave it several more powerful rips with his jaws, enlarging the hole just enough. Then, putting his nose in first, he bellied through with such force that the jagged ends of wood cut deeply into his neck. But he was through.

Inside, great choking walls of smoke and flame reared up. The once-familiar utility room had become a grotesque inferno. King raced toward the hall leading to Howard and Fern's room, but the fire pushed him back. It scorched the fur on his head and shoulders and scalded the soft pads on his feet.

Again King leaped into the wall of flame; again he was pushed back. But there, just there, an opening! Quickly he took his chance and bounded through the slim break that led in the only possible direction – toward 16-year-old Pearl's room.

The girl was in a deep sleep. King whined anxiously, putting his singed head on her arm and nudging her with his nose. In a semi-conscious state Pearl shoved at King and rolled over. Urgently he tugged harder, and finally she sat up in alarm.

Now, dog and girl groped their way to a second door leading to her parents' room. When Pearl shook her mother awake, the smoke was so thick Fern needed no explanation. She cried out to Howard and, choking and unable to see, stumbled with Pearl to the bedroom window, and escape, believing her husband would follow.

Once Fern and Pearl were outside, King could hear them calling his name through the thick soup of smoke. But, when Fern tried to get him to jump, he headed back into the murk. He had to let her know that Howard, overcome by the smoke and his already-bad lungs, hadn't gotten up from the bed.

In pain, tormented by his burning paws and the gash in his neck, King returned to Howard's side. Weak and confused, his master got up but only to stumble in the wrong direction and fall inertly to the floor in the hall. Frantically King ran back to the window, and this time Fern understood. She climbed back inside and let the big dog lead her to her stricken husband.

King stood still while Fern roused Howard, waiting all the precious minutes it took him to get to his feet and put one arm around her. Then, with Howard's other hand gripped tightly to the fur on King's neck, they led the half-conscious man through the searing heat and smoke to safety.

Out in the cold night air at last, King hovered protectively next to the elder Carlsons while Pearl ran to the neighbors for help. Not until the neighbors arrived and the wail of sirens bore down upon the gutted house did he relax his vigil. And it wasn't until the revealing light of dawn illuminated the destruction caused by the fire that the Carlsons realized just how much their courageous pet had risked for them.

King had a mouth full of painful splinters, an ugly wound on his neck, singed hair the length of his body and paws so badly blistered they would be overly sensitive for a year.

For his unflinching devotion and courage in saving the lives of Howard, Fern and Pearl Carlson, the indomitable King was awarded the Ken-L Ration Dog Hero Award for 1981. His feat of tearing through the plywood door was considered so amazing he was featured on the television show *That's Incredible*.

Fawn

Fawn heard the muffled buzz, dry and sharp. Sprawled on the floor of the screened porch, the German shepherd cocked the big ears that had earned her her name, lifted her shoulder off the wooden flooring, and listened intently. Nearby, her owners, Mr. and Mrs. Schlesinger, and their married daughter, Sue Schoenberger, continued to visit, unaware of the noise. Only Fawn recognized the rattlesnake's angry warning.

The four-year-old, 75-pound shepherd was a favorite among the Schlesingers' four dogs. Always protective, she usually accompanied Mr. Schlesinger, who ran a one-man auto supply firm in St. Petersberg, Florida, on his delivery rounds. Sitting up front in the big van, tall ears alert, Fawn would caution people away from his supplies. And, only a few months earlier, she'd scared off a pair of would-be burglars who were attempting to chisel the lock off the Schlesingers' front door.

Now, on this October day in 1974, she was up immediately and out the dog port cut in the screen. The rattler was not alone on the lawn. Playing nearby was three-year-old Russ Schoenberger, nicknamed "Tiger" by his grandparents. From the porch steps, Fawn saw Russ treading through the grass only inches from the four-foot diamondback. The big snake was coiled and poised to strike, its back studded with sun-struck scales.

Fawn growled a warning and charged. Russ toddled on, oblivious, toward the deadly snake. The snake's sleek head froze, zeroing in on its target.

Fawn bowled into Russ, shoving him sideways and knocking the bewildered boy to the ground just as the rattler's wide-open mouth slashed through the air where he'd been an instant before. Turning in a flash, Fawn lunged, snarling, toward the toddler's assailant. The snake fixed on the dog and a ribbon of tongue cruised from its mouth. Its long body swayed. Arching back in a giant question mark, the rattler again poised to strike, this time at Fawn.

Fawn dodged the attack and lunged again, coming in close enough to nip at the snake's side. But the deadly, poisonous mouth was close to her own and Fawn pulled back. Then Russ was standing up in the grass, still only inches away from the fray. The snake drew up to strike in his direction.

Never taking her eyes off that lethal mouth, Fawn intervened again, knocking the little boy down with her rump as she circled the deadly diamondback. In the same move, she butted the rattler sideways with her nose, always careful to keep herself between Russ and the venomous reptile.

Crying and confused, little Russ reeled to his feet once more. Trying to bypass Fawn, the snake slithered quickly toward the boy's pudgy legs. Fawn grabbed for the snake while simultaneously flinging Russ aside once again. The toddler howled with indignant anger.

This time, however, Fawn lost the dangerous balancing act. As the Schlesingers came on the run, they saw their dog lunge. Her teeth were headed for the snake's fat, meaty body. But the rattler lashed out, its head streaking arrow-straight and its needle-sharp fangs sinking into the soft flesh of the dog's lip.

Russ ran to the safety of his mother's arms while Mr. Schlesinger, now armed with a .32-caliber pistol, called to Fawn. At last the wounded dog relinquished her stance and staggered back. Then, after a series of shots which ended the threat to Russ and the other family members, she sank onto the grass, her face already puffing painfully.

It was an anxious 20-minute ride to the nearest veterinarian. Fawn's face swelled gruesomely and her breathing narrowed to a whistle. Ten more minutes and the shepherd would not have survived. As it was, her face remained swollen for three weeks, aflame with the venom meant for three-year-old Russ. Little "Tiger" would have had even less chance for recovery.

Boldy risking her own life for Russ Schoenberger, the big-eared dog was named Dog Hero of the Year by Ken-L Ration. Attending the celebration in Chicago with Fawn, the small friend whose life she had saved summed it all up by clapping his hands and proclaiming, "I'm a lucky duck."

Priscilla

Lake Somerville felt cool to Priscilla, providing welcome relief from the oppressive heat of the Texas sun that late afternoon of July 29, 1984. At two months old, the 22-pound piglet had a natural love of water and had been swimming almost since birth in pools around her home in San Marcos.

This, however, was a real treat – a day at the lake with Carol Burk and her 11-year-old son, Anthony. For hours she'd delighted Anthony, who was mentally impaired and couldn't swim, with her water antics. Now she and Carol had just gone out for one last plunge in the lake, with Carol holding the long leash attached to Priscilla's harness as the young, black-and-white pig kicked her already powerful legs in steady rhythm.

Priscilla loved people. Her owner, Ada Davis, had weaned her on a bottle after her natural mother had rejected her, and Priscilla had grown accustomed to the playpen on Ada's porch and to Ada's gentle, motherly ways. Already she felt different from the other pigs on the farm. Priscilla couldn't bring herself to eat with them, and wasn't much interested in the pig games that might get her pink ears and snout dirty. Ada had ministered to the piglet as tenderly as she might her own child, and perhaps Priscilla felt she was.

Priscilla was already slightly winded from swimming circles around Carol when she heard her friend's startled shout. They were a long way from shore and could barely touch bottom. But it wasn't Carol who was in trouble. It was Anthony! Oblivious to his mother's frantic warnings, the boy was up to his waist in water and moving quickly toward them.

Priscilla saw the trusting grin on Anthony's face. She saw Carol waving urgently at him to go back, and she saw Anthony's boyishly thin arms reaching forward as if to pull the water to him, still unmindful of any danger. Then, suddenly, Anthony hit a deep spot and went under.

Crying out, Carol dropped Priscilla's leash and struggled in the direction where her son went down. But she was agonizingly far away. She watched helplessly as the water churned with the boy's desperate, awkward attempts to surface. For a moment only his small hand appeared, clutching at nothing.

Priscilla had been swimming in a widening arc near Carol. Alarmed, she now changed course and bore down upon Anthony. She swam quickly, forgetting that she was tired, legs working earnestly and round snout snorting with effort. Anthony had been under water for a dangerously long time.

Suddenly the boy's head popped up. His face was stark with panic. Flailing helplessly he gulped and spluttered. Then he choked on swallowed water and disappeared again.

Carol was still too far away to help, but Priscilla was almost on the drowning boy. Urgently she worked to close the gap. Anthony was nearly gone when she reached him. Priscilla swam close, and instinctively the boy grabbed for her leash. But, in panic and exhaustion, he clawed too desperately, and Priscilla felt herself being pulled under.

They were both sinking now. Anthony weighed almost four times as much as Priscilla, but she knew how to swim, and he was holding tight to the leash. Priscilla knew they needed air. Again and again she twisted in the frothing water, resisting his panicky flailing. The glimmer of sunlight on top of the water seemed hopelessly far away.

Priscilla's lungs were bursting; her legs growing weaker and weaker. If she couldn't pull them up now they were both going to drown! With one last enormous heave, the little pig righted herself and headed for daylight once more.

This time, when Anthony surfaced, he seemed to grasp his rescuer's intentions. Too weary to resist, he clung to the leash while Priscilla swam. Priscilla, who'd already expended incredible energy for a pig her size, was exhausted too. But she headed stubbornly for shore, not stopping until Anthony could once again touch bottom. Carol reached them moments later.

Priscilla had saved Anthony's life.

For her heroics the young pig received the William O. Stillman Award for 1984. Priscilla accepted the award in style, dressed in matching purple cape and bikini. Her home state of Texas proved so charmed by her courage and devotion that she became the first pig ever inducted into the Texas Pet Hall of Fame.

For her unswerving bravery in saving Anthony's life, Houston Mayor Kathryn Whitmire proclaimed Saturday, August 25, 1984, to be "Priscilla the Pig Day."

Priscilla

Panda

Fresh dew soaked Panda's paws as she sat in the thick grass watching her mistress, Mrs. John Traines, hike through her Lopez, Washington, pasture. It was an early spring morning in 1942, and Mrs. Traines had begun checking her stock, which included several Jersey cows and a bull.

Before entering the pasture, Mrs. Traines had told Panda, a black, shepherd/bird dog mix, to wait behind the fence. Just one-and-a-half years old, Panda hated any separation from her mistress, and she squirmed restlessly as she saw the trail of Mrs. Traines' heavy boots lengthening through the grass.

The huge bull, who'd been lying down with his hooves tucked beneath him in a noncommital pose, fixed his eyes on Mrs. Traines with a glassy stare. Although she was a good distance from him, he snorted angrily and rose to his feet. Lowering his big head, he pawed the ground and let out a warning bellow.

Alarmed by this sudden show of aggression, Mrs. Traines turned away from the massive beast and headed back toward the fence, which lay several yards to her right. The bull bellowed a second time and lumbered in her direction.

Mrs. Traines stopped. The bull stopped too, churning up the ground with his hooves. He threw his head back defiantly. Then he stood perfectly still for a moment, as if sizing up the situation. In the next second he took a few backward steps, lowered his head — and charged.

Knowing she was trapped, Mrs. Traines turned and ran for her life. It wasn't much of a race. She was too far from the fence, and, in a few quick steps, the surprisingly agile bull had closed the gap between them. Within seconds the helpless woman was inches away from being trampled and gored.

Panda, hearing the bull's first bellow, forgot her mistress' order to stay out of the pasture and wriggled under the fence. Sensing the bull's belligerence, she stood rigid, watching the drama unfold as Mrs. Traines tried to avoid a confrontation with the maddened animal. Then the situation turned ugly.

As the bull bellowed again and charged, Panda shot across the pasture, her eyes riveted on the bull's galloping bulk. Running as fast as she could, the dog skimmed along the ground, intent on intercepting the big brute before he reached Mrs. Traines.

In spite of Panda's lightning speed, the bull was nearly on top of the fleeing woman when the dog shot between them. Frantic to save her mistress, the courageous canine slipped on the slick grass but came up with an angry growl. Gathering herself, she sailed through the air, landing with her teeth clamped deeply into the bull's one vulnerable spot — his tender nose.

Infuriated, the bull grunted and ceased his headlong charge. Jolting to a stop, he threw his head back violently, whipping Panda skyward. Blue sky, clouds and trees whirled past, but when the bull's neck snapped forward Panda was still hanging on. With shuddering force, the bull slammed the dog into the ground, and Panda yelped in pain as her leg was fractured.

The crazed bovine bellowed in rage, but Panda refused to let go. She kept her teeth locked tightly into the bull's nose. The bull slammed her downward again, cracking several of her ribs with a shattering blow. Still Panda hung on.

Meanwhile, the dog's fierce onslaught was holding the bull at bay, buying Mrs. Traines enough time to reach the safety of the fence line. As she crawled through the protective wires, the bull continued whipping Panda violently from side to side.

Panda yelped again, and Mrs. Traines called desperately to her from behind the fence. Hearing her mistress's voice, the wounded dog released her hold. Instantly the bull swiped viciously with his horns, laying open a deep wound in Panda's side. The dog rolled away and stood up.

Mrs. Traines called again, and Panda staggered toward the sound of her voice, narrowly avoiding a second swipe from the deadly horns. The weary dog managed to dodge still another attack, and to crawl just out of the mad bull's reach. She edged toward the fence and, once under the wire, collapsed at Mrs. Traines' feet.

Mrs. Traines was not hurt from her encounter with the bull, but the injuries Panda received during the battle to save her mistress proved too numerous and serious for the young shepherd/bird dog. She had paid a terrible price for saving Mrs. Traines. Panda died that same night.

In recognition of the courage the young mixed-breed dog had displayed in sacrificing her life for Mrs. Traines, Panda was nominated posthumously for the Latham Foundation Gold Medal Award for Heroism.

Fido

Fido heard the crack of the bat and raced after the ball sailing through the evening air. The excited children screeched instructions to the small brown-and-white mongrel as he raced after it.

Skipping across two front yards and into a third, the mutt turned abruptly and, after letting the ball bounce once or twice toward him, caught it in his teeth — and the children cheered.

For seven years Fido had been a popular playmate in the neighborhood games. Belonging to 10-year-old Laurel Schall, he tagged along as a welcome member of the Linden Hills, Minnesota, gang. That brisk April evening in 1931 was no exception.

A few minutes later Laurel suddenly remembered her promise to be in before dark. She grabbed up her baseball mitt and called to Fido. Then she turned and started for Linden Boulevard, the large street she had to cross to reach home.

In a rush, she dashed across the grass toward the busy street. As she reached it, one of her playmates called good-bye, and the little girl turned to wave, forgetting to look where she was going as she stepped off the curb. At that moment, a heavy sedan was closing rapidly on the spot.

Fido, on the lawn several yards away from Laurel, saw the car bearing down on his mistress. With a loud bark he leaped after her, running harder than he ever had before.

Hurtling the last few feet, Fido slammed into Laurel's midriff, knocking her sideways and spinning her out of harm's way. The speeding car missed Laurel by inches, but not Fido. The skidding tires crushed the life out of the brave little dog.

Laurel and the other neighborhood children had lost a friend. The loveable tag-along they'd all taken for granted had given his life to save one of theirs. In mourning their pal's death, they nominated him for the Latham Foundation's Gold Medal Award for heroism.

Chips

When Chips, a tough-looking combination German shepherd/Husky, arrived at the War Dog Training Center in Front Royal, Virginia, no one would have guessed just what an impression he would eventually make. In time, Chips would become the first and only dog ever to be nominated for the Silver Star for gallantry in action.

Chips, donated to the war effort by a family in Pleasantville, New York, was a member of the first War Dog detachment sent overseas from the United States. The year was 1942, and the world was involved in a war of staggering proportions – with the outcome still in question.

In training Chips quickly proved to have the ability to respond to commands both correctly and with a special brand of ingenuity. He quickly mastered the important skills of a scout and sentry, both jobs he would be called upon to perform when he went to war.

After training and transfer overseas, Chips reported for duty with the 30th Infantry, 3rd Infantry Division. Within days Chips began to show the mettle of a soldier with a promising future. Hitting the beaches of North Africa with his division in November, 1942, Chips immediately became useful as a sentry with keen sight and powerful jaws.

Following their invasion of North Africa, Allied soldiers were being bothered by roving bands of Senegalese natives, who would sneak into the camps at night to steal clothing and equipment. Chips and other dogs quickly went to work on the problem, catching a number of thieves and holding them for the patrolling guards. The natives found the dogs to be strong, silent and ferocious fighters. The random forays into the camp quickly came to a halt.

The future hero of the battlefield was also to brush against world leaders as the war went on. His first such encounter occurred in January, 1943, when he was assigned to help guard the house where President Franklin Roosevelt and Prime Minister Winston Churchill met for the historic Casablanca conference. At the conclusion of the conference Roosevelt and Churchill took the time to meet Chips and review his fellow protectors.

Within months the American army was on the move again, heading toward Sicily. The campaign to take Sicily began with a huge amphibious assault. Landing in the pre-dawn darkness with his handler, Private John R. Rowell, Chips and the others were quickly pinned to the beach by a hail of machinegun fire from an enemy bunker. With the Allied forces virtually trapped between the machinegun nest and the sea, Chips took matters into his own paws.

Leaping from Private Rowell's grip, Chips charged the bunker. Sand spit in every direction as the enemy fired at the enraged beast heading straight for them. First, a bullet nicked Chips in the scalp, and then a bullet tore into his hip. Still Chips charged, now more angry than ever. An instant later the canine soldier tore into the bunker, and all that could be heard on the beach was the savage clawing of flesh and clothing, and the screams of those inside. Suddenly, a soldier staggered out – with Chips solidly clamped to his neck. Next, three more men came out with hands raised, pleading to surrender. Chips had captured the machinegun nest, and had undoubtedly saved the lives of many men on the beach.

Major General Lucian K. Truscott, Jr., Commander of the 3rd Infantry Division, acted on recommendation to award Chips a Silver Star for bravery, plus a Purple Heart for wounds received in action. Such an award was unprecedented in military history.

On November 19, 1943, Chips stood in a church courtyard in Pientravairano, Italy, to receive his medals. Reading from the citation, the Division General spoke of a special brand of courage, arising from love of master and duty. "Chips' courageous act," the General read, "single-handedly eliminating a dangerous machinegun nest, reflects the highest credit on himself and the military service." Chips' medal was withdrawn, however, following protests by those who felt that it was inappropriate to award an animal a decoration meant for human heroism.

Chips slowly recovered from his gunshot wounds and burns in his mouth from grabbing the hot barrel of the machinegun. Eventually he caught up with the Division and continued in combat through Italy and on to France, Germany and Austria. Chips maintained his reputation of fearlessness, and only once can it be said that his instincts ever failed him. On that occasion General Dwight Eisenhower had stopped by the 3rd Division for a visit. Not in full military dress, "Ike," as he was known, bent over to pet the famous nominee for the Silver Star. Chips, trained to be wary of strangers making unexpected moves, bit his hand. Surely Eisenhower understood.

When the war in Europe ended, Chips was sent home to Front Royal and discharged on December 10, 1945. The decorated war hero was unofficially awarded the Theater Ribbon by his unit, along with a Battle Star for each of the eight campaigns in which he was involved.

Chips, the dog hero, was then returned to his original owners in New York to live out his life.

Chips

Pal

Pal, a mule owned by J.E. Campbell of Prineville, Oregon, had an unusual habit. He liked to follow his master around his cattle ranch.

Unbeknownst to Campbell, the mule trailed after him one spring day in 1931 as the 60-year-old rancher strode through the pasture. A range bull had broken the fence and entered the field where some young Jerseys were grazing, and Campbell, spry for his years, was determined to chase him out.

The sagacious Pal was one of a team of mules Campbell often used on the ranch. Muscular and sure-footed, his propensity for seeking human affection had earned him his name. His big ears twitched expectantly now as he let himself through the broken wire behind Campbell, listening to the confident tone the elderly man adopted as he approached the bull.

The great animal, unaccustomed to having his authority challenged, swung his head toward Campbell, his wide-set eyes glaring. He stomped and snorted, sending streamers of drool floating out of his mouth in the breeze. Campbell took another step forward.

Suddenly the unpredictable beast lowered his head, scooped his horns through the air – and charged. The ground shook with a thunderous pounding as he headed straight for the defenseless rancher. Realizing his error in an instant, Campbell looked quickly around for some kind of weapon. There was only bare, uneven pasture. Without hesitating, he turned and ran.

Active as he was, the aging man was no match for the bull. The maddened bovine bellowed furiously at his retreating form and raced after him. Campbell glanced over his shoulder and saw the bull closing the gap between them. He looked ahead to the fence line, still many yards away. He doubted he could make it.

Then, without warning, Campbell tripped and fell. As the ground rushed up to meet him, he knew he was done for. He'd been a rancher for too long not to have heard the stories of bull-gored victims. Their fate was often a severe mauling – or worse.

There were only a few seconds left before the bull would reach him. Automatically, Campbell crossed his arms in front of his face. In that same instant, however, he saw a flash of brown streak by and gaped in amazement.

Charging with all the stature and fleetness of his female parent, and the stubborness of his sire, Pal was heading directly for the oncoming bull. Head down, ears flattened backward, he moved past Campbell in a blur.

Pal and the bull met a few feet from the prostrate man. In a single swift motion Pal swung around. Avoiding the lowered horns, he sent his hind hooves shooting into the bull's rib cage.

The bull, outweighing Pal by hundreds of pounds, lumbered to a surprised stop. He sent forth a furious bellow across the pasture. Pal didn't wait for the horns to come slashing; he whirled a second time and kicked the bewildered bull in the ribs again.

Campbell, taking advantage of Pal's intervention, scrambled to his feet while the mule held the bull off. He made for the fence and a spot where, earlier, he'd seen a large stick. Gaining the stick, he wheeled and ran back. He knew that Pal, outmatched as he was, couldn't last much longer.

By now the mule had his opponent fully engaged. The bull's massive head and shoulders swayed threateningly as Pal kept just out of reach of the horns. But he was tiring. Rivulets of dirty sweat ran down his neck and shoulders. His hindquarters trembled with effort. The bull, unfatigued, stood solid as a wall of bricks.

Campbell called to Pal and advanced on the bull, brandishing the makeshift club. Hearing him, the bull snorted defiantly. Not liking the looks of the club, and finding himself outnumbered, the big brute turned tail at last, allowing Campbell to drive him from the pasture.

Pal's devotion to Campbell had saved the rancher's life, and both man and mule came out of the ordeal unharmed. For his uncanny intelligence and swift action, the mule was awarded the Latham Foundation's Gold Medal for the State of Oregon, receiving the honor in 1931.

Afterwards he became what he'd always wanted to be: the pampered pet of the Campbell ranch.

Trixie & Duke

The sun was setting across the trout stream in the hills above Bakersfield, California, as H.E. Hobright waded out to cast his fly. From a distance, the water looked placid but, standing hip deep, the fisherman could feel the strong tug of the current as the stream's cold waters came down from the mountains that spring evening in 1930. It sashayed past his chest-high waders and tugged relentlessly at his feet, which he'd planted on mossy rocks in the stream bed.

While Hobright fished, Trixie and Duke, his two dogs, poked among the boulders on the bank. Trixie was a small fox terrier and Duke a tall, loose-jointed hound. They were as unlike as dogs could be.

Mischievous Trixie was marked by a white coat and two black ears, with a pirate's patch over one eye. Sleek-haired, fatherly looking Duke boasted giant feet, a long, swinging tail and a curiously whiskered nose. Together they were a Mutt and Jeff duo, with the tiny, light-on-her-feet Trixie often winding around beneath Duke's lanky legs with ease.

As the sun skimmed the horizon to the west, Hobright caught the flash of a rainbow trout working the opposite bank and drew out line. Upstream to his right was a series of huge boulders jutting into the water. He wanted to reach that fish and, while the current was hard to battle, he thought that if he could make the boulders he'd find some protection and a handhold there. Cautiously he went deeper into the stream.

As he neared the boulders, his foot caught in a crevice between two rocks, twisting his ankle beneath him. Crying out, he lunged forward, striking his head on granite as he fell. His rod went sailing out of his hand, and water quickly filled his waders. Acting like an anchor, the heavy rubber boots dragged him under.

On the bank, Duke and Trixie both looked up in alarm as Hobright cried out. Without warning, their big, six-foot master was unconscious in the water, his face submerged and fast disappearing.

Duke bounded to the edge of the stream and jumped into the icy flow after Hobright. Kicking out, he swam in an arc below the drowning man, trying to head up-current toward him. As he neared him, the swift maelstrom pulled them apart. Duke had to circle again, this time coming close enough to seize his master's fly vest in his teeth.

Duke pulled and headed for shore, but the combination of filled waders and the draw of the stream gave him no slack. He'd kept Hobright from floating off downstream, but he couldn't seem to drag him to the bank.

Floundering, the lanky dog circled in the stream, not knowing what to do. Taking hold of Hobright's vest again, he snorted with the enormous effort it took just to keep them in one place. In a few more minutes, there would be no hope for the half-submerged man.

Trixie, watching Duke's helpless maneuvers from the bank, ran along the rocky edge as if debating whether or not to join him. But if Duke couldn't budge the unconscious Hobright, there was little hope that she could.

Suddenly the small dog stopped pacing. Instead of heading into the stream, she turned and shot up the steep trail which led to the road.

Up and up the patch-eyed dog ran, dodging the scrape of low branches and scrambling over boulders many times her size. It was a long, quarter-mile trek to the top. Several times her short legs buckled beneath her and she skittered downhill in the shale. Each time, Trixie valiantly picked herself up and raced ahead again.

At last she reached the top of the incline and flat ground. Here she skipped into the middle of the country road and barked urgently. A lone fisherman, Mr. A. Brown of Bakersfield, heard the incredible distress in her voice as he hiked nearby and decided to investigate.

Approaching the little dog slowly, Brown knelt and called to her. But Trixie was dancing excitedly on the asphalt and wouldn't be calmed. When he stood up she immediately raced back to the steep trail, barking over her shoulder.

Curious, Brown peered over the incline. Through the trees he saw Duke, still struggling valiantly in the stream to keep Hobright from washing away.

Quickly Brown slid down the trail after Trixie and eased carefully into the swift water. As he grabbed Hobright's arm he felt, with surprise, the leaden weight of the man. "I'll still need you," he said to Duke. He didn't know if the dog understood or not, but it didn't matter – Duke wasn't about to let go of Hobright.

Lunging out, the exhausted dog swam on, tugging at Hobright's fly vest. Now, with Brown pulling, the two were able to beat the current and bring the unconscious man to shore.

H.E. Hobright recovered from his near-drowning and a nasty gash on the head. Among the headlines of the day praising the amazing duo of Trixie and Duke, his rescuers, was one from *The San Francisco Examiner.* It hailed the "quick thinking and perfect teamwork" which enabled the two dogs to join forces and save their master's life. Further recognition came from the Latham Foundation with awards for animal heroism.

Dolphin

Because much is still unknown about cetacean behavior, experts often disagree about the validity of acts of altruism performed by the dolphin or porpoise. Yet the fact remains that, among seafaring people throughout history, there has always been an abundance of stories acknowledging the friendliness and concern these highly intelligent, warm-blooded creatures have shown toward their human counterparts. Legends and true accounts of the helpfulness and courage dolphins and porpoises have displayed in saving human lives have come to us from around the world, most notably from the Greeks, Romans and Polynesians.

The earliest of these stories comes from about 600 B.C. when the lyre-playing Greek, Arion, was said to have been robbed by pirates and thrown overboard while enroute to Cornith. A wild dolphin or porpoise (names of two slightly different species, often used interchangeably) was credited with carrying the drowning man to safety on its back.

Similarly, ancient legend records that in the Roman colonial town of Hippo on the northern coast of Africa, a dolphin was said to have saved a boy from drowning when the youngster had swum too far out to sea. Maneuvering beneath the boy and taking him on its back, the dolphin carried the astonished child to shore, returning for months afterward to amuse its new-found friend and his playmates.

In modern times, evidence that dolphins might delight in – and even seek out – the company of humans is best expressed in stories from New Zealand describing the exploits of Pelorus Jack and Opo.

Pelorus Jack, as one of these dolphins came to be known, spent 24 years, from 1888 onward, "guiding" ships through French Pass, a dangerous channel off New Zealand's coast, and into Pelorus Sound. So grateful were seafarers of the region to the rare Risso's dolphin for saving them from the harrowing reefs and whirlpools of French Pass that he became the first of his kind to be legally protected from harm through a law signed by the Governor of New Zealand in 1904.

New Zealand's second famous dolphin, Opo, appeared offshore in 1955 to awe the inhabitants of Opononi with her antics. Opo "played tag" with youngsters swimming in the shallows, allowing herself to be stroked and petted, and even gave children rides on her back.

More recently, dolphins have reportedly driven sharks from swimmers and guided fog-bound fishermen to safety. The life of Bob Marx, a New Zealand treasure hunter, was saved in November, 1984, when two dolphins drove away an attacking hammerhead shark; and, in 1982, friendly dolphins drove sharks from the water around New Zealand's marathon relay swimmer, John Curry.

Another recent account, coming from South Africa in 1978, relates how a boat filled with fishermen, stranded by fog off a dangerous coast, was nudged through a series of treacherous shoals by a school of dolphins. Their unselfish concern for the fishermen earned the dolphins a high regard from their land-dwelling neighbors.

The following story, gleaned from U.S. newspapers, is an account of a dolphin who saved the life of a woman on the coast of Florida in 1941. Reporters filing the account wrote that the woman wished to have her name withheld.

It was a sultry Florida day. The beach was nearly deserted and the surf seemed calm. Discarding her sandals, the woman padded over the wet sand to the water's edge. Waves rushed across her feet and receded in hissing foam. She waded out, unconcerned.

The undertow grabbed her when she was only waist deep. In a split second she was dragged under, water pouring into her nose and mouth, cascading down her throat into her lungs. In a shower of bubbles she was flipped head-over-heels, her right cheek hitting hard against the sea's scalloped bottom. Spewed up moments later by the surging surf, she struggled against the current and called out weakly. She was being drawn into deep water.

Not far away a dolphin surfaced, the blowhole on top of its head widening as it expelled and then sucked in air. Then it turned toward shore to satisfy its curiosity about the presence of the woman in the surf. At the same time, it noticed that a shark was also closing in on the helpless swimmer being held captive by the ocean's undertow.

Leaping once, it dove and bore down on the predatory fish. The dolphin had superior speed and intelligence over its hated enemy. The shark's gaping mouth revealed rows of razor-sharp teeth and, while one shark by itself was not necessarily a threat to the dolphin, blood streaming from flesh torn by those vicious teeth would bring others of its kind in from miles around for the kill.

With swift deliberation, the dolphin aimed its torpedo-like snout for the big fish's most vulnerable area – its gills. Moving at full speed, it raced through the water, slamming into those vital slits just behind the shark's head. The impact stunned the fish, and the dolphin didn't wait for it to retaliate. A second thrust sent the shark moving out to sea for deeper water.

With another slap of its tail the dolphin sped toward the woman, who was now floating face-down and barely moving. It surfaced near her, its great liquid eyes staring at the woman's limp form as it circled her warily. Then it dove.

Maneuvering beneath the woman, the 250-pound mammal rose to the surface, lifting the woman's face and upper body out of the water. With the

Dolphin

next forward surge of the waves, the dolphin moved toward the beach, the woman draped across its glistening back.

As the dolphin neared the beach it whipped in an abrupt turn, propelling the woman off its back and into shallow water. The woman, on her hands and knees, gasped and coughed, shaken, but alive.

It was several minutes before the woman could recover enough to rise, turn and try to thank the individual who had rescued her. She assumed that another person had jumped into the swirling surf and pushed her to safety.

In fact, a man had come over the dunes during her ordeal, but he had been too far away to be of assistance, only reaching her at the water's edge.

What the still-astonished man saw, the woman saw when she turned her gaze back out over the breakers: The dorsal fin of a large shark knifing through the water and out to sea. Behind the shark, herding it away from the beach and leaping toward the horizon, was the silvery form of the sleek, life-saving, dolphin.

Caesar

I

Caesar paced the sidewalk in San Francisco's cool night air, listening for the one car he could distinguish above the din of all the others. It would be the Studebaker belonging to his master, swimming coach Phillip Patterson.

A popular athlete, Patterson worked at the San Francisco Hotel plunge, and Caesar, his long-legged German shepherd, eagerly awaited his arrival home each night. That fall evening in 1931, the alert dog heard the familiar engine's whine and wagged his tail sharply as the Studebaker rounded the last corner, heading for the Roosevelt Street garage.

Just as the car slowed for the ramp leading into the building, a menacing shadow leaped from behind a wall. The shadow materialized as a burly man in a dark mask, furtively running alongside the Studebaker. Suddenly the man jumped onto the runningboard and, hanging his right arm through the car's open window, thrust a gun at Patterson.

Patterson swerved, narrowly missing a block column, and braked to a halt in the dark garage. Trapped in the car with the loaded weapon threatening at point-blank range, the coach sat helpless.

Gruffly the man demanded money, his dark, jittery eyes darting over Patterson's face. When Patterson hesitated, the man pushed the gun's barrel against his temple. The assailant's voice cracked with the dangerous, crazed rasp, and Patterson realized with dread that, no matter how he responded, the man meant to shoot him.

The moment Caesar saw the gunman jump on the runningboard, he had left his usual greeting place and scrambled after Patterson's car. As he crept up behind the Studebaker, he could hear the shadowy man shouting at his master. Softly he padded along the concrete floor until he was only a few feet from Patterson and the armed thief. The man's voice had risen to a frantic dangerous pitch.

Crouching low, Ceasar sprang toward the stranger, knocking him away from the car and sinking his teeth into the calf muscle of the man's leg. Startled, the man fired at his unknown attacker, but the bullet ricocheted through the darkness, barely missing Caesar.

The man screamed and began clubbing Caesar with the gun butt. Several hard blows crashed on Caesar's head and neck, but his teeth stayed locked on the gunman's leg. His angry snarls echoed through the cavernous garage as the sound of the shot died away.

Patterson, with the gun no longer trained on him, quickly ducked down in the car, anticipating more gunfire. Meanwhile, the thief thrashed about wildly, finally beating the dog off with a hard kick to the chest. Caesar sprang away as the desperate man came to his feet. The gun barrel was now pointing at the shepherd's head.

Teeth bared, Caesar sprang again. The second shot exploded in the air, whistling past his ear. It had missed its mark. Quickly the dog toppled the gunman, sending his weapon clattering out of reach.

Unarmed, the would-be burglar scrambled to his feet and took off running, Caesar in hot pursuit. The shepherd chased the man out of sight and then returned home, weary but without serious injury.

Still shaken, Patterson emerged from his car and ran to summon help. Police arriving on the scene agreed that the gutsy shepherd had saved the coach from being robbed, and possibly killed, by his armed assailant.

Other than a few nasty bruises, Patterson and his courageous dog were unharmed after the attack in the Roosevelt Street garage. For saving his master's life, Caesar received the Latham Foundation's Bronze Medal in 1932.

Bill Dog

Long after the others in the Eberhardt's Sonoma, California, home were asleep, Bill Dog was kept awake by the pain of his amputated leg. Well into that April, 1930, night he fidgeted and moaned weakly on his square of rug.

Of an indeterminate breed, Bill Dog's tenacity had been tested to the limit for the past two weeks. Out chasing rabbits early one morning, he had been caught in a steel trap. The cruel-jawed snare had kept him pinned and bleeding for 14 days before the Eberhardts finally found him.

Nearly dead from exposure and lack of food, Bill Dog stayed alive by drinking rainwater that collected in a puddle near his head. The damage to his leg, however, proved irreversible. After a few days of careful nursing by Mrs. Eberhardt, the infected limb had to be removed.

That was yesterday, and tonight the postsurgery shock and the anesthetic retreating from Bill Dog's system left him drained and aching. He hardly had the strength to respond when he smelled the fire.

It began in the kitchen of the family's home, the ominous smoke seeping thickly through the rooms ahead of the deadly flames. Sleeping soundly, neither Mr. nor Mrs. Eberhardt, nor any of their children, were aroused by the impending disaster.

Alarmed, Bill Dog tried to lift himself from his rug but quickly found the source of his pain. Where once he had stood on four legs, now there were only three. Scrambling for balance on the newly bandaged stump, he fell clumsily to his belly.

Whimpering, Bill Dog rolled over on his good side. He tried to bark a warning but his throat, raw from days spent hoping to attract attention while in the trap, only emitted a pitiful howl. His feeble cries went unheeded.

The wounded dog tried to rise again. It took him several attempts, his stump throbbing whenever he moved. Fresh blood flowered onto the bandages. But, with great effort, he found he could make slow progress by using a hopping gait, relying more heavily on his front legs than he ever had before.

His rug had been laid out in the bathroom, and it was from here that he began the stubborn, awkward hopping that would carry him to Mrs. Eberhardt's bedroom. Each step was a milestone. Bill Dog moved as if the two halves of his body, front and back, had been disconnected. Throwing his weight forward on his front legs he'd pause, waiting for his stump and remaining hind leg to catch up. Behind him he could hear an ominous crackle as the fire drew near. The growing heat pressed against his coat.

Slowly, painfully, Bill Dog arrived at Mrs. Eberhardt's door. Pushing it open, he could see her sleeping form under the covers in a series of pocketed shadows. Already under the deadly grip of the heavy smoke, she coughed fitfully but didn't wake. Bill Dog lurched forward, determined to rouse her.

A short distance away he paused, crouched, and clumsily jumped toward the bed. The leap wasn't a good one. As soon as he landed, only halfway on the bed, he began to slide backward. He sank his teeth into the covers, hung on, and kicked forward again.

Mrs. Eberhardt, as groggy as she had become, was startled awake when Bill Dog landed on top of her, licking her face and whining. It had taken him a quarter of an hour to crawl to her bedside. A few more minutes and no one would have left the house alive.

Bolting upright in surprise at Bill Dog's sudden arrival, Mrs. Eberhardt blinked into the haze and smelled the fire. Grabbing the courageous canine, she pulled herself from the room, shouting in alarm. The smoke was so poisonous she was forced to sink to her hands and knees to keep from passing out. Through the mounting heat she struggled to take over Bill Dog's job to rouse the sleeping family.

Finally, it seemed to Mrs. Eberhardt that her husband and children were all accounted for, and together the half-conscious family members groped their way out of the burning house. Once outside, still confused and shaken, they stared back at the inferno they'd barely escaped. Flames poured from the windows and sent sparks wavering into the night sky. Mr. Eberhardt put Bill Dog down and collapsed into her husband's arms.

It was then that the injured dog, frantically looking around, let forth his pitiful howls once again. The Eberhardts, in their muddled state, had failed to realize that their youngest daughter hadn't made it out with the others.

Suddenly Mrs. Eberhardt understood, and she turned to her husband in horror. Fighting the flames, they dashed back inside and found their daughter by her bed where she'd fallen, unconscious. Scooping her into his arms, Mr. Eberhardt rushed from the house with his wife as the fire totally consumed their home behind them.

Bill Dog's love and detemination saved his family's lives and earned him a nomination for the Latham Foundation's Gold Medal Award.

Hero

Dainty-footed and narrow-faced, Hero was a purebred collie and a show dog – the pride of the Jolley family's farm near Priest River, Idaho. It was here, in the rural setting he loved, that he kept alert to the rhythms of the farm and the daily routine of the people and livestock living there.

One late afternoon in 1966, the farm's activities were highlighted by the horses coming in from pasture. Skittish and tense, they thundered past him on their way toward the barn, raising clouds of Idaho dust. Hero expertly guided their trek, coaxing any strays back into line with a series of quick movements performed at their heels.

Arriving at the barn, Hero stayed outside while the big animals headed for their stalls. Heavy hooves drummed the packed earth as they jostled for position. Impatiently, they stomped and snorted, nostrils flaring. Mrs. Jolley, her three-year-old son, Shawn, at her side, began pitching hay down to them from the barn's loft.

Suddenly the familiar routine was shattered by a shrill scream. Hero became instantly alert. In spite of the din of the horses, he had recognized young Shawn's voice. The collie was already streaking in that direction when he heard a second shout, unmistakable this time. It was Mrs. Jolley calling to Hero in alarm.

He ran for the open door. Inside the air was hazy, streaked with sunbeams streaming through the cracks in the barn's board siding. Mrs. Jolley, high in the loft, stood paralyzed, too far away to reach the scene unfolding before her. Shawn, no longer at her side, ran terrified across the floor below, an enraged stallion at his heels.

Somehow, the little boy had descended from the loft unnoticed by his mother. Now, with the moody chemistry of stallions, the big horse had turned on him. There was no doubt that the angered beast would trample him to death.

Defenseless, Shawn was trying to reach the shelter of a space beneath a parked tractor. But, as he dove for cover, his jacket caught on a projecting piece of metal and he was pinned there, fully exposed to the horse's fury. The stallion screamed and reared up, flashing hooves descending toward the helpless boy.

At that moment Hero leaped between them, flinging himself at the maddened horse and locking his jaws on the vulnerable skin of the stallion's nose. For a few tense moments horse and dog were locked in combat. Hero was lifted high in the air and shaken furiously. Blood and spittle flew. The stallion's deafening screams filled the air.

109

Then Hero was wrenched from side to side with sickening force and flung through the air toward one of the tractor's wheels. The collie was slammed mercilessly into the unyielding tire and pain seared through his chest. Broken ribs made his breath come sharply. Shawn still lay trapped before the horse, his small, grubby face wild with fear. The murderous horse, blood streaking from the teeth wounds in his muzzle, reared up once more.

Hero was up in an instant. Again he planted himself between Shawn and certain death, taking the stomping meant for the little boy. The horse was 15 times Hero's size, and his sharp, deadly hooves came crashing down on the collie's forefeet, crushing them. Hero yelped in agony but would not give ground.

Now, out of the corner of his eye, he could see Mrs. Jolley descending from the loft. She needed time to untangle Shawn's jacket and shove him to safety under the tractor. Because of his broken ribs, every breath Hero drew was filled with needles of pain. His crushed forepaws responded like hot, weighty clubs. Yet he again lunged at his huge opponent. This time the stallion's striking hooves glanced off his jaw, the force of the blow knocking out several teeth.

Mrs. Jolley was given the precious seconds she needed. At last Shawn was safely in the recess beneath the tractor. Now the stallion turned his maddened violence solely on Hero.

Powerful hooves sliced viciously through the air, continually pounding the exhausted and badly injured dog. Hero never faltered. He matched the crazed stallion's merciless attack with equal fury, snapping and barking, weaving in and out to avoid the lethal blows.

Mrs. Jolley joined the battle, picking up a stick and thrusting it at the horse, trying to keep him from killing Hero. Then, just as suddenly as the attack began, it ended. The horse stood still, chest heaving, with blood and saliva splotched crazily across his neck and shoulders. With a last angry shake of his head, he snorted and headed for the open barn door, Hero still snapping at his heels.

Every step was torture for the embattled collie, but Hero gave chase until his adversary was out of the barn and galloping across a distant pasture. Only then did the brave dog acknowledge his severe wounds, dropping to the ground with blood spewing out of his nose and mouth. He was rushed to a veterinarian some 45 miles from the Jolleys' farm and, amazingly, he recovered from his wounds in only six short weeks.

The valiant dog was hailed as having saved three-year-old Shawn Jolley's life. For his amazing stamina and courage in facing the raging stallion, Hero received Ken-L Ration's Dog Hero of the Year Award for 1966, being the first "show dog" ever to capture such an honor.

Hero

Baby

Baby lifted her feline nose and worriedly sniffed the air. In the dark interior of Minnie and Ray King's Piney Flats, Tennessee, home, the gray-and-white cat sprang to her feet. She smelled smoke!

It was 4:30 a.m. on January 20, 1984. Baby had been a member of the household since four years before, when the Kings had found her, pregnant and sick, on their doorstep. She was about to repay them for their kindness.

Baby raced back down the hall. Outside her family's bedroom door she halted, meowing loudly, sending distressed cat calls into the quiet room. Her cries echoed plaintively along the walls.

In the bedroom, Minnie King slowly aroused from a deep sleep. Hearing Baby's cries, she called out in a scolding voice of her own, saying something about Baby's breakfast having to wait a few more hours. Then she rolled over and drifted back toward her dreams.

Frantic, Baby rushed into the room and leaped on the bed. Now she set up such a yowl that both Kings came awake in earnest. What had gotten into their cat?

Suddenly they heard a loud crash coming from somewhere in the basement. Minnie clutched Ray's arm in fright. They faced each other in the darkened bedroom and tried to hush Baby. It sounded to them as if an intruder had invaded their home. Cautiously Ray pulled himself from the bed and padded down the hall, following the sound.

What met him when he opened the laundry room door threatened Ray King more than any thief. The unchecked fire roared forward, and a great column of smoke pushed him backwards. Realizing that their entire home would soon be in flames, Ray yelled to his wife and raced for the phone. One call to the fire department and hastily thrown on clothes were all they had time for.

Once in the safety of the street, Ray and Minnie King stood watching the smoke billow from the fire. They'd still be in there if Baby hadn't awakened them.

Baby! Only then did Minnie realize that, in their haste to leave the inferno, they had not seen Baby follow them out. Minnie made a quick decision and made a determined dash back into the burning building, calling her cat.

Shut up in the house, Baby was pacing the smoke-filled rooms. Already the floor was scorching under her feet, and the poisonous air choked her when she breathed. She'd tried the usual exits through both doors, finding them shut tight against her. There had been nothing to do but wait.

At last Baby heard her mistress call. The gray-and-white cat scrambled toward her in the heavy air and Minnie bent down, scooping her up in her arms. Together the woman and cat fled the flames, which roared through the entire house only moments after their escape.

Baby, the brave cat, had sounded the alarm that saved her adopted family's lives. Her swift action earned her the Latham Foundation's Gold Medal Award for heroic animals.

Prince

It was December 16, 1984, and traffic on Long Island's Hempstead Turnpike thinned to a trickle as midnight approached. It had been a quiet evening in East Meadow, where Noebeck's gas station was located.

Roger, the station's young attendant, was glad for the company of Prince, the station owner's five-year-old German shepherd. Roger had good reason to appreciate Prince's presence. In his short lifetime, the vigilant canine had thwarted no less than 10 robbery attempts. He knew how to do his job, which was to protect his master's property and the men who worked there.

The crash of breaking glass split the stillness of the night. On the far side of the building, three men were entering a side window. They wore ski masks and gloves. In their hands they carried knives, a saw, a heavy pickaxe and a seven-foot crowbar.

As the men grabbed Roger, Prince lunged, sinking his teeth into the arm of one of the robbers. The man swung the crowbar, clubbing Prince in the neck. But Prince hung on and the man went down, Prince on top of him.

The second man, armed with the pickaxe, swung wildly and drove the point of the heavy tool into Prince's body. As the dog spun around in pain, the first man repeatedly bludgeoned his head with the crowbar. The odds were obviously against him, but Prince did not hesitate in his attack.

The German shepherd went for the nearest leg, bringing down the third masked man, and hung on while waves of pain shot through his head. The man with the pickaxe swung again, striking Prince in the throat, but Prince kept his teeth locked on the third man's leg.

A new shock hit him as a knife plunged into his body and entered his stomach. The pain was blinding, but Prince fought the oncoming darkness. His throat and stomach wounds gushed blood on the concrete floor as he reeled, fell, and momentarily lost consciousness.

Coming to, Prince realized that Roger had been bound with rope, and the three men had pried open the floor safe. They were leaving with his master's money. With his throat cut and his stomach split wide open, Prince made a last, vain attempt to stop the thieves. His wounds were too severe, however, and he could move only a few short feet before he again collapsed, unconscious, his muscles twitching from spasms of pain.

As soon as the men were gone, Roger rolled out the door and into the street, stopping the first car passing the station. Help was finally on the way.

When the police arrived, however, Prince was once again alert, but his mind was seared with pain and anger. No one he didn't know was going to enter the station. He growled fiercely at any new intruder, snapping his teeth when Officer Miller tried to approach. For 20 minutes Prince continued to hold the police at bay, and only when the dart from the tranquilizer gun stung his skin did he relent. The warm glow of oblivion flooded over him, releasing him from the agony this terrible night had brought.

Prince was rushed to the Westbury Animal Hospital, but he had lost too much blood. He died on the operating table, but he was not forgotten.

Praised in newspapers of New York and beyond the next day, Prince was accorded the recognition of a true hero.

Grizzly Bear

Grizzly Bear, an enormous but extremely amiable Saint Bernard, lived with Mr. and Mrs. David Gratias, the owners of a rustic lodge in Denali, Alaska.

One cold spring day, when the noon light was falling through the windows of the Gratias' cabin, Mrs. Gratias heard noises outside. Investigating, she left the cabin door open behind her. Her two-year-old daughter, Theresa, was sleeping, and she wanted to be able to hear if the child awakened.

She paused to scratch Grizzly Bear's drooping ears. The dog trotted after her, nose raised for the sweet smell of new spring grass.

In the backyard, Mrs. Gratias discovered a grizzly cub. When she saw the little brownish-yellow bear, it didn't look at all dangerous, sitting dolefully on its fat rump, but Mrs. Gratias was immediately apprehensive. She knew that, where there is a bear cub, there is certain to be a fiercely protective mother bear nearby.

Except in winter, grizzlies have no fixed homes and roam over vast territories, eating everything from grass to caribou and an occasional moose. Although ordinarily shy and reclusive, a grizzly protecting its young can be vicious and unpredictable. Mrs. Gratias suddenly remembered the open door and her small daughter napping just inside. Alarmed, she turned and raced back around the cabin.

As she rounded the corner, there was the enormous mother grizzly. The huge creature bellowed and raised herself up to her full eight-foot height, slashing at Mrs. Gratias. The woman tried to side-step, but slipped and went down.

Dazed by her fall, she was only dimly aware of the grizzly scraping her cheek or of its great curved claws gouging like chisels deeply into her shoulder. She was aware of the bear bent over her, jaws agape, ready to inflict a fatal bite, but there was nothing she could do to defend herself. She caught the glint of enraged eyes descending... when abruptly the bear pitched backwards.

Grizzly Bear, the Saint Bernard, had smashed into the huge bear's chest with all of his 180 pounds. The bear regained her balance ungracefully, bawling with rage. Again she went for Mrs. Gratias, but Grizzly Bear placed himself between the beast and the helpless woman. With a growl the big dog charged again into the bear with teeth and claws.

Mrs. Gratias was losing blood rapidly from her wounds. Her growing weakness mingled with her shock and terror as she lapsed in and out of consciousness. The snarling Saint Bernard squared off to protect her.

117

The mother bear moved forward ominously on her short hind legs. Suddenly she lunged and swiped at the powerful dog. Grizzly Bear dodged quickly, but felt the three-inch claws comb through his long coat. Fearlessly he retaliated, his teeth gripping the bear's tender black nose. The bear's savage eyes widened in shock. Roaring defiantly, she rose on her hind legs again, her ripped snout dripping blood.

The bear's massive head weaved as she lunged forward again. Even on all fours she was almost twice as tall at the shoulders as the Saint Bernard. Grizzly Bear backed off, turning the bear away from Mrs. Gratias.

The she-bear ignored the woman now; she was after the red-and-white dog. Her powerful jaws were open, displaying a bone-crushing array of teeth. Warily, she rose on her hind legs again, then sank to all fours. Grizzly Bear saw his chance and dove in for another attack, striking at her eyes and face, then leaping back before she could seize him.

The monster's fury was murderous now. Her loose skin and shaggy coat shook hysterically. She charged the dog, her usual shuffling gait disappearing in a blur of savage velocity.

Grizzly Bear dodged, then charged once more, sinking his teeth deeply into the bear's bleeding nose.

The bear's momentum collapsed. She snapped her head back. Grizzly Bear released his grip and moved out of reach again. The bear turned a full circle, shaking her head and bellowing. A chunk of her tender nose was missing now, and through the pain and bewilderment she suddenly seemed to remember what she was fighting about. Her eyes searched for her offspring, but the cub had long ago scuttled off to safety. The bear turned back to the dog.

Grizzly Bear stood stiff-legged near his fallen mistress. His lips were lifted back in a snarl. Finally the she-bear backed off, cautiously, treading heavily on her big hind paws. Defeated and whimpering, she turned and trotted away.

Grizzly Bear leaned over Mrs. Gratias. He sniffed and nuzzled her gently, then tentatively licked her bleeding cheek.

She roused and sat up. She was dazed, her mind trailing the wisp of some bad dream. Suddenly, she remembered her daughter, struggled to stand, and hurried to the door of the cabin. Grizzly Bear followed. They found little Theresa still sleeping soundly.

Mrs. Gratias' wounds healed. And miraculously, Grizzly Bear, in his battle with one of the largest and most ferocious carnivores on Earth, had emerged without a scratch. For his great courage and loyalty, he became the Ken-L Ration Dog Hero of the Year for 1970.

Grizzly Bear

Bo

Water from the Colorado River fanned in rainbow-hued arcs from the sides of Rob and Laurie Roberts' rubber raft and sprayed high against the cobalt sky. The Glenwood Springs couple was shooting the rapids on a clear spring day in 1982. With Rob and Laurie in the little craft were Bo, the Roberts' Labrador retriever, and their new pup, Duchess.

Bo, his dark coat dripping, sat alert in the quivering raft's rubber bottom as the upsurge crashed in drumming sheets against its taut skin. Suddenly the water narrowed as the river ran through a boulder-choked channel and shot up in booming froth. Speeding through the neck, the raft shuddered and spun, the incredible force of the river throwing it out of control. The little group watched in terror as a tall wave rose up from the maelstrom.

The wave grew to eight feet, its menacing face obliterating the deep blue sky. For a perilous second it hung suspended, then came crashing down on the raft. Bo was on all fours as the craft upended and shot crazily skyward.

There was a huge sucking noise above the roar of the rapids when the raft rose up from the water, then flipped completely over. Rob and Duchess were flung sideways, out of harm's way, but the small boat crashed down on Laurie and Bo, pinning them under the unyielding rubber.

Swept far downstream, Rob and Duchess made their way to shore. From there Rob waited anxiously, unable to fight the powerful downstream drag to rescue Laurie. In agony, he watched as the seconds ticked by and the raging water did not release his wife.

Trapped beneath the smothering rubber, pummeled by the river's icy currents, the dog and his mistress were struggling for their lives. Boulders scraped their limbs and the mad rush of the water pulled them along, smothering their cries of distress in its thunderous crescendo.

Finally, Bo managed to wriggle free of the deadly snare and surface in the torrent. Not Laurie. No matter how she strained, the raft remained clamped over her like a tight-fitting lid.

Bo spluttered and gulped air, trying to hold his position in the freezing water. While pinned beneath the raft he'd been aware of Laurie beside him, and now, with a few hasty kicks, he rejected the safety of the shore and returned to the capsized craft. Diving beneath the surface, he again battled the powerful downstream pressure, heading straight for his drowning mistress.

Laurie, quickly exhausted from her fight for air, now thrashed feebly in her underwater prison. Bo found her grappling form and sunk his teeth into the first thing he could get hold of – a mass of splayed hair. Gripping tightly, he tugged her down, away from the suction of the rubber, before turning abruptly and heading up.

Surfacing together, the shivering pair bobbed in the violent froth. The few minutes in the river had brought them both dangerously close to drowning, and a tired Bo circled Laurie as she coughed and sucked in oxygen. Then he swam by her side and the trembling woman reached out, grabbing her dog's tail and letting him pull her to shore.

To the reunited Roberts, it was evident that Bo had saved Laurie from a tragic death. For his courageous rescue, the brave Lab was honored with the Ken-L Ration Dog Hero of the Year Award.

Gero

Police Officer Mike Pruitt of the Gainesville, Florida, police department and his partner, a six-year-old German shepherd named Gero, liked to reward themselves with a stop at a local convenience store after a good day's work. Their purchase was always a candy bar, which they shared. But on one spring day in 1986, as they responded to a burglary call, their small, symbolic moment of friendship was to be pushed many hours into the future.

Arriving at the scene of the burglary, they found that other officers had set up a "perimeter" at the building on N.E. First Street. Gero scrambled from the patrol car, following his partner, and together they took cover. There was the possibility that the armed and surrounded thief would choose to shoot it out.

Gero, whose Indian name meant "he who walks in his shadow," had been on the K-9 force for four years. Good training had taught him what his duties were as Pruitt's "shadow" when the desperate thief chose to dash from the building, gun raised.

As the thief charged the police blockade, lowered his gun and shot at the crouching officers, Gero rushed him. Panicked, he wheeled and fired again, blasting the hurtling Gero through the shoulder.

The heavy-caliber slug did not stop the dog's lunge. As blood spurted from a neat hole in his coat, Gero flew at the thief and locked his teeth on the gunman's arm. In the next few seconds the thief grappled with the growling shepherd, his fingers still clutching his weapon. Another shot split the air as he fired a third time, sending a bullet into Gero at point-blank range.

The force of the blast flung Gero back harshly, coloring his fur with powder burns and blood. The sky wavered and the hard sidewalk slammed up to meet the stunned dog. Quickly the thief turned his gun on the police officers, sending bullets screaming over their heads as they ducked for cover.

More shots rang out as the officers returned fire. Rushing forward, Officer Pruitt found himself momentarily exposed, feeling a brief surge of panic as he saw the thief leveling the deadly gun on him. He knew he had but moments to live if the gunman fired.

Gero, though terribly wounded, was aware. The thief's weapon, which had inflicted his own terrible pain, was aimed at the man he'd ridden with and grown to love over the past many months.

In a single, torturous lunge, Gero's body became a shield between Pruitt and the gunman. Gero took the bullet himself.

The dog's courageous move gave Pruitt and another officer just enough time. Firing two shots each, they dropped the thief where he stood. Rushing to the friend and partner who'd saved his life, Pruitt found Gero's once-proud body already curled in death.

At the memorial service honoring the K-9 hero who'd acted far above and beyond the call of duty, more than 100 of his fellow officers watched as Officer Pruitt placed an unwrapped candy bar in the short, stout coffin.

The Thoroughbred Messenger

They chose the best horse they had, Colonel Henry Carrington's tall Thoroughbred. He was brought stomping from his Fort Phil Kearney stall in Wyoming Territory that December 21, 1866. Hands numbed from the frigid 25°-below weather brushed him and placed the heavy saddle across his back. A compact, dark-skinned rider settled into the leather, and the rough-hewn stockade gates were unlocked. The bars slipped open as the pair shot into the midnight bleakness.

The well-bundled rider, John "Portugee" Phillips, and the prancing stallion left behind 119 men, along with their women and children, dependent upon the outcome of their ride. Fort Phil Kearney had been the scene of a violent massacre, and now the survivors were surrounded by 3,000 warring Sioux. So desperate were the trapped whites that the fuse was already laid to blow the powder in the magazine, where the women and children would wait out the last stand.

The fort's tiny band of defenders knew there was a 100-to-1 chance that the dispatch held close to Phillips' breast would get through to save them.

Outside the gate a fresh blizzard blowing in off the Bighorn Mountains slashed at horse and rider. The Thoroughbred shivered but needed no prodding. Given rein, he lunged southeast over the snow-buried sage toward the Bozeman Trail. Their final destination was Laramie, 236 miles away.

Phillips sank low across the Thoroughbred's neck and laid his hand on the Sharps carbine scabbarded beneath his leg. The eerie cry of the Sioux had whispered toward them on the wind. The Thoroughbred's powerful stride lengthened and Phillips sent up a prayer that the darkness would hide them.

Through the night the Thoroughbred battled the storm and the sage-choked wilds, guided only by Phillips' sense of direction. At daybreak the pair hid themselves in a gulch. Here Phillips pulled the blanket from the big horse every few minutes and marched him up and down to keep them both from freezing.

It was an anxious, sleepless day. By nightfall Phillips' arms felt leaden through his buffalo coat. His chest sharp-pained with cold. Foam flecked the corners of the fatigued Thoroughbred's mouth and turned into icicles on his whiskers. Saddling up, the weary pair headed out of the gulch toward their first stop, Fort Reno.

But just as they were about to clear the gulch, a nervous flick of his mount's ears made Phillips turn in his tracks. A Sioux warrior was outlined on the rise behind them.

Phillips ducked as the brave sent a bullet slamming past his head. A second brave and then a third appeared on the rise. Spurs dug into the Thoroughbred's sides as he plowed through the snow-drifted gulch. Halfway up the incline the big horse lunged and stumbled backwards, while from the opposite side the Sioux charged down toward them.

The Thoroughbred let out a snort and plunged forward through the hip-deep snow. In a matter of minutes the Indian ponies would be on them. The Colonel's horse took the challenge and bounded up the incline, clearing the top with seconds to spare. Once on level ground the Kentucky stallion shrilled defiantly, racing away from the ponies with the speed born of his breed.

On through the darkness the Thoroughbred ran, through Crazy Woman's Fork of the Powder, a notorious setting for past scalpings, toward the square-topped rocks of Pumpkin Buttes. It was sometime in the middle of the night when they made Fort Reno. Phillips called to the sentry on duty to open the gates, made it inside the stockade and slumped, spent and drained, from the saddle. He patted the drooping horse in admiration. They'd covered a distance of 70 miles.

Hot coffee for the rider and a rubdown for the horse awaited them as Phillips relayed his urgent message to the Commandant. Outside the wind howled as the anguished man told Phillips he hadn't a soldier to spare. Phillips nodded solemnly, swallowed the rest of his coffee, and returned to the Colonel's horse.

The Thoroughbred's ribs and hipbones had begun to show beneath the leather, but he stamped his hooves and threw his neck high at Phillips' approach. The rider swung into the saddle and together the pair raced back into the winter darkness.

Once again they stopped near daybreak, camping under a cutback. Phillips removed the saddle and rubbed the Thoroughbred down. The horse's muzzle was crusty with ice, his legs bore the countless scratches of the windswept sage, and his withers trembled under Phillips' touch. It would be another sleepless day on guard for Sioux and the ever-present fear of freezing.

Not until night fell did Phillips look up into the clearing sky with sudden realization. It was Christmas Eve.

They spent the holiday creeping through the slicing wind, trying to keep on course and stay awake. Walking, trotting, galloping, walking again. Hooves crunching through the powdery snow to rocks and sage, floundering in drifts; cold, stunning to the bones. Phillips' eyes were gritty and dry as he locked them on the landscape for any hostile signs.

It was 10 o'clock Christmas morning when the Thoroughbred at last stumbled along the telegraph line into the town of Horseshoe, another 160 miles from Fort Phil Kearney.

"Good God, they're nearly frozen!" the shouts went up. Concerned hands pulled Phillips from the saddle and examined the Colonel's horse. Their message was taken immediately to the telegraph operator, who sent it out over the wires. But there was no answering tap from Laramie, 40 miles away.

It was a thin, sallow-eyed Phillips who mounted the gaunt and flagging stallion one last time. Flanked by two men from Horseshoe, with the message

Thoroughbred Messenger

once more against his breast, the nearly prostrate but determined Portugee bent and spoke in the horse's ear.

"We're taking it on in," he said. Then he gave rein and the intrepid Thoroughbred trudged forward, even as the second wave of sleet and snow descended across the prairie.

This time they rode in daylight, but the dense gray swirl suggested dusk. Countless times the Thoroughbred and the other two horses floundered in belly-high drifts. Lanced by the north wind, the stallion's once-silky neck bent low. The snow had obliterated the Overland Trail, and Phillips headed them in what he hoped was southeast, toward the regiment at Laramie.

Night fell. The gray world turned black: a treacherous maw that urged them to halt, to rest and sleep where they stood. Maybe they had missed the town altogether. Maybe their message would come too late to save the survivors of Fort Phil Kearney.

Yet the Thoroughbred plodded on.

At last the Colonel's horse raised his head and Phillips peered through the sleet. Lights! And the vague shape of buildings. They'd made it. Phillips placed a grateful hand on the nearly comatose steed.

The General was attending a dance. The bright notes of a song tinkled out into the night. The low current of celebrating voices could be heard. The big horse stumbled toward the sound. Past the parade ground, the barracks and stores. Right up to the veranda steps. Phillips slid rigidly from the saddle and staggered up the stairs toward the summoned General. His hands shook violently as he pulled the dispatch from his coat.

As soon as he delivered his message, the exhausted Phillips fell incoherent on the floor. He was to live, but outside in the snow the mighty Thoroughbred pitched forward, dead.

Years later a marker placed by the State of Wyoming would mark the spot where the gallant horse had dropped. A horse who endured against impossible odds to save more than 100 lives.

Dutch

As the shaggy German shepherd paused on the shore, listening, feathery tufts of hair along his ears and legs danced in the cold breeze. Dutch, as the big dog was called, followed the antics of two boys with lively interest as they tussled on a small pier near Troy, Pennsylvania, their raucous shouts bouncing across the freezing pond's water.

Dutch's young companions, Gordon and Hugh Hawthorne, were a boisterous, fun-loving pair. At three and four years old, respectively, they played and fought together with comradely zeal. On that day in 1963, however, their playful mischief was to turn deadly serious.

It was little Gordon who fell in the pond, stumbling on the rough-planked pier and arcing into the air, away from his brother. There was a resounding smack as he hit the 34-degree water. From the dock, Hugh watched his struggling brother with momentary disbelief, seeing Gordon's pinched, gasping face pleading to him from below. Already the sinking boy's eyes had lost their brightness and were clouding over with panic. The pond was 12 feet deep, and Gordon couldn't swim a stroke.

Neither could Hugh, but that made little difference to the horrified four-year-old. He only knew that his brother was drowning. Yelling to Gordon that he was coming, Hugh jumped off the edge of the dock.

Gasping, Hugh popped up near his brother's submerging form. Gordon was writhing wildly as he sank beneath the water, his contorted face turning ivory. Hugh reached out for him, even as he felt himself going down as well, helpless in the water's frigid grasp. Like his brother, Hugh flailed his arms and kicked with his feet in a furious dance to find something to grasp or stand on. With a swirl of bubbles, the water closed relentlessly over his head.

At that instant a dark, shaggy form knifed through the bubbles and surged powerfully toward the boys. It was Dutch, swimming for all he was worth to reach Gordon and Hugh before they sank from sight.

Reaching Hugh first, Dutch swam close to the frightened boy, but Hugh's struggle for survival overrode comprehension and he kicked out fiercely, catching the shepherd in the chest. Undeterred, Dutch circled him, his limbs growing numb in the glacial pond. In moments Hugh grew exhausted and his thrashing began to ebb, his lungs filling with icy water as he sank toward the pond's murky bottom.

With a powerful effort, Dutch dove beneath the surface, following Hugh. The depths were dark and confusing, but the shepherd moved swiftly. Lunging through the half-frozen muck he gripped the boy's ankle in his jaws, turned and headed for daylight.

Quickly Dutch surfaced with Hugh in tow. Never loosening his grip, he plunged awkwardly toward shore, the weight of his burden threatening to pull them both back under. Kicking and straining, he at last delivered the spluttering boy to dry land.

Dutch stood over the quaking Hugh and barked urgently at Mrs. Hawthorne's approach. The boys' mother, nine months pregnant, had first been alerted to her sons' plight by the dog's urgent cries as he leaped into the pond. Walking quickly, then running toward the commotion as fast as her swollen body would allow, she reached the shore soon after Dutch emerged with Hugh. Ordering the shepherd to stay, she dove for Gordon, who was floating face down and seemingly lifeless.

While Dutch licked and nuzzled the sobbing Hugh, Mrs. Hawthorne swam hurriedly to her youngest son. When she turned him over his face was a delicate blue, and his half-opened eyes stared at nothing. Appalled, she pulled him to shore.

Carrying Gordon from the water, Mrs. Hawthorne worked on his inert form for what seemed like hours, pushing the water from his lungs. At last she heard an answering gurgle, and rushed to the telephone for help.

According to police officers and doctors who arrived at the scene to revive him, the three-year-old was "technically dead" three times before he was successfully brought back to life at Troy Memorial Hospital. A few moments more in the freezing water and their efforts would have been in vain.

By sounding the alarm and acting quickly, the shaggy, spunky Dutch had saved the lives of the two boys. Later that year, fully recovered and boisterous as ever, Gordon and Hugh Hawthorne were on hand to applaud their pet's heroics. For his double rescue, Dutch was awarded Ken-L Ration's Dog Hero of the Year Award for 1963.

Ringo

It was difficult to decide exactly what breed of dog Ringo was. He had the large frame of a Saint Bernard, the short, thick fur of a malemute, and the high, alert ears of a German shepherd. From all appearances, he looked as if he had been assembled by the proverbial committee.

Trotting down the streets of Euless, Texas, beside his mistress' little two-and-a-half-year-old son, Randy Saleh, the big dog was both imposing and comical by turns. Clearly the most impressive thing about Ringo that day in 1968 was his determination not to let Randy out of his sight.

A compulsive wanderer, Randy had slipped away from his mother's watchful eye only minutes before construction was to begin on a fence she'd planned to build to halt his adventures. Ignorant of any danger, the child had been exploring the streets of Euless for two jubilant hours, even as his frantic mother and the local police searched for him in vain.

Now Randy was getting tired. He'd covered nearly three-quarters of a mile in the sultry Texas air, sidetracking whenever something caught his eye. His wanderings had brought them right to the edge of busy Pipeline Road. Beside him, Ringo paused when he did, glancing worriedly ahead. Hot gusts of dirty exhaust flew in his face and pressed back his chest ruff as a steady stream of cars barreled past. They were facing a notoriously dangerous blind curve at the bottom of a hill.

For a few moments Randy watched, fascinated, as the cars appeared around the curve and shot along the roadway, their swift, enormous wheels flashing by in a blur. Finally the hiss of approaching tires died down for a moment, and the little boy stepped nonchalantly onto the asphalt.

Randy reached the middle of the road with Ringo walking protectively at his side. Here was something new. He squatted on a divider line and ran his stubby fingers along the smooth paint, delighted with its feel. He paid no attention to the sound of another car descending the hill beyond the curve, the whine of its engine racing toward them.

Alarmed, Ringo bounded forward around the curve, barking and snarling at the oncoming car. The driver saw the dog and honked her horn frantically, but Ringo wouldn't budge. Not only wouldn't he budge, but he was leaping toward her, teeth barred. The motorist swerved and braked hard, squealing to a stop only inches in front of him.

With the car halted, Ringo raced back over the hill and around the curve to where Randy still played in the roadway unseen. With his big nose, he nudged the toddler off the asphalt. But no sooner had he gotten him to safety than Randy turned and headed out again, delighted with the new game.

Ringo had no time to urge him back once more. He had heard another car flying down the hill behind him. Quickly he whirled and raced around the curve.

The first motorist had just begun to ease her car forward when Ringo again came charging toward her. This time she felt the thud of the dog's heavy shoulder against her fender. He'd leaped at her car as though to attack it.

Unnerved, the motorist threw her car into neutral and put her hand to her hammering heart. It was obvious the dog was completely insane.

Snarling a warning, Ringo then raced behind the first car to halt the next car, which he did by again placing himself in the roadway. The second motorist, seeing the stalled car and the bristling dog, screeched to a stop.

Having succeeded in holding them off momentarily, Ringo then dashed out of sight to where Randy sat waiting on the painted lines. With anxious prodding, he directed the boy off the road a second time.

Now, in the distance, Ringo could hear a third car speeding down the hill. Even as he raced to block it, Randy toddled gleefully into the roadway yet again. The siege had begun.

Five cars, ten cars, and then twenty. As each vehicle descended the hill it was met by Ringo and the increasing snarl of traffic. The big dog, frothy-mouthed with fatigue, ran blindly back and forth between his mechanical adversaries and his little charge, bristling at one and coddling the other. The rumor of "mad dog!" raced down the line of imprisoned motorists, and no one dared try to approach him.

Several minutes later Harley Jones, a school maintenance employee, counted 40 backed-up motorists when he was brought to a stop at the end of the line. Deciding to investigate, he learned that the earliest captives had been held there for a quarter of an hour. In awe, he watched as the exhausted Ringo still thrust himself with wild fury against the veritable army of cars, attacking any bumper that tried to roll past him and growling sharp warnings at opening doors.

Giving the dog a wide berth, Harley walked cautiously around the stalled cars and rounded the curve. What he saw there amazed him even more. Little Randy was plopped, laughing, in the center of the road.

At that moment Ringo came rushing to the boy and again began coaxing him from his lethal spot. Realizing the situation at once, Harley eased closer, talking soothingly to the big, exhausted dog. For a moment Ringo stood perplexed at this turn of events, calming enough for Harley to reach Randy and swing him into his arms.

Quickly Ringo was at the stranger, his long teeth inches from Harley's heels as the man carried Randy to the side of the road. But, once he'd gained it and the cars began to file slowly past, Ringo relaxed at last, his tense vigilance over.

By using his own body as a shield, Ringo had set up a miraculous barricade on Pipeline Road – a barricade that had saved young Randy Saleh from certain death in the onrushing traffic. That same year, 1968, the "mad dog" was honored with the Ken-L Ration Dog Hero Award for his extraordinary devotion.

Caesar

II

In November, 1943, the first Marine Corps dog unit entered the fight against the Japanese during a beach assault on Bougainville in the Solomon Islands.

Attached to the 2nd Marine Raider Regiment (Provisional), the 24-dog unit conducted itself courageously in battle, thereby proving to everyone that dogs were valuable wartime comrades.

One of the Marine War Dogs stood out from all the rest, however, and won the hearts of his fellow soldiers thoughout the combat zone and the corps.

Day One of the assault found Caesar, a three-year-old German shepherd, with his handler, PFC Rufus Mayo, on the forward edge of M Company's position near the jungle. The Japanese, meanwhile, were using the jungle to the best possible military advantage. With dense growth as a shield, the enemy were keeping the Marines bogged down.

Along a narrow trail called "Foxhole Avenue," Mayo and Caesar awaited instructions. When communications between M Company and the Second Battalion were disrupted, it became necessary to physically carry dispatches between those in command. Caesar was selected as the messenger and served as the only link between those on the front line and the command group in the rear. It was extremely hazardous duty, punctuated by explosions and gunfire – some aimed directly at Caesar. His mission was critical. Marine lives might be saved if only he could continue his runs successfully.

On the second day of the battle, communication with the front was briefly re-established. But when the lines were once more cut by the enemy, the big German shepherd was again called upon to run the trail with vital messages.

Caesar, working day and night, eventually made nine runs between M Company's position and the rear command post. Twice he came under heavy enemy fire as the Japanese, well aware of the dog's critical mission, attempted to shoot the burly canine and isolate the Marines. Despite the bullets whizzing by his ears and tearing up the dirt around his fast-moving paws, Caesar never faltered.

The Marines were amazed at the dog's stamina and unerring sense of duty. Caesar never became lost in the jungle or strayed from his mission. In constant danger, he repeatedly traversed the deadly corridor. The Marines, especially in the forward area, began to believe Caesar was invincible.

The battle raged into the third day as the Marines fought to drive the Japanese forces back. For Caesar, the two previous days had been grueling, yet the hard work had paid off. Communication was no longer a problem. Caesar, the hero, might have expected a night of rest as reward. Unfortunately, the foxhole he occupied with PFC Mayo was several hundred yards forward of the main company. Their mission was to act as a listening post in case the enemy launched a surprise night attack.

Mayo and Caesar were hunched in the foxhole when the shepherd's head suddenly cocked ever so slightly. A chill went down Mayo's spine. The young Marine realized the dog heard something in the jungle; a terrifying experience at night, when any noise might mean instant death. Caesar had discovered an enemy soldier creeping silently through the jungle, and now rushed to attack.

Instantly Caesar and the Japanese soldier were pitted in a deadly struggle. The dog clenched the soldier's arm in his massive jaws, inflicting terrible damage. In the midst of the fight, however, the soldier somehow fired twice. Then all was silent. PFC Mayo could make out the shadow of the enemy running away, most likely taking with him an advancing party of Japanese soldiers.

Mayo anxiously scanned the jungle beyond his foxhole for his companion. Then, a few feet away, he saw Caesar struggling toward him, gravely injured. Caesar's coat was soaked with blood; both of the enemy's bullets had hit their target. Mayo held Caesar and yelled for a medic.

Caesar was placed on a stretcher and moved to the rear. The Japanese, meanwhile, had learned a valuable lesson about sneak attacks at night; with dogs on duty, the advantage was always with the Marines.

At first light Mayo checked the area and found the spot where Caesar had fought with the Japanese soldier. On the ground was an unexploded hand grenade, undoubtedly intended for Mayo's foxhole. Caesar had again saved lives, and soon Marines in every foxhole and forward position were inquiring about his condition.

Caesar, although severely wounded, recovered from his injuries and was given a special commendation for services rendered on Bougainville. In a personal letter to Max Glazer of the Bronx, New York, who had donated Caesar to the Marines, Commandant of the Marine Corps, General Thomas Holcomb, credited Caesar for "saving the lives of many Marines."

Caesar, the German shepherd from the Bronx, had proved his courage and loyalty in the service of his fellow Marines. To them, he was a comrade – a soldier who shared every hardship and terror of war – and who, in the most harrowing of circumstances, performed with magnificent heroism.

Caesar

Anonymous

The little Boston bulldog trotted behind 10-year-old Jack Sigmon as he jogged through the Oklahoma woods near Anadarko one summer day in 1926. Jack and his family, originally from Cushing, were staying in the country for the summer. The bulldog was on loan to them from Roy McKee of Anadarko. The stout, friendly little animal stayed close to the heels of the young boy, following him when he left the house with his familiar cap squashed down on his head, to round up the horses.

They found the small herd grazing in a secluded pasture nearly a half-mile from the house. Jack ordered the bulldog to wait under a tree while he approached the lead mare. She was a tall, cantankerous sorrel, and he walked toward her slowly, halter ready and calling softly as he went.

The sorrel eyed Jack suspiciously and threw her head high, making it impossible for him to slip the strap over her ears. He tried again. This time the mare side-stepped and snorted. Then, without warning, she wheeled on him and lashed out.

Jack was knocked to the grass and saw, in an instant of terror, the heavy hooves descending. There was a bone-chilling crack as they met his skull. Stunned by the lightning attack, the boy lay semiconscious on the uneven ground.

The bulldog was up immediately. With the speed and agility born of generations of his ancestors, the broad-chested, muscular-backed dog raced toward the big sorrel. Thrusting his wide, black nose in the air, he barked shrilly, succeeding in driving the horse from Jack's inert form.

As the horse galloped away, the dog whined and bent over Jack. Jack tried to move but the world was spinning. A trickle of blood came from both ears.

Planting his short, stout legs in the grass, the bulldog sank his teeth into Jack's shirt collar and tugged, trying to pull him back to the house. Somewhere through the reeling in his brain the boy understood and saw himself reaching out to touch the dog, even crawling as he was being urged to do. Still he lay unmoving on the ground.

The bulldog licked the boy's face, sniffing at him worriedly. He took hold of the collar a second time, managing to drag Jack a few inches. Jack groaned and passed out.

For some time, the bulldog continued to circle Jack, repeatedly latching onto his shirt and pulling with all his might. It was no use. Despite the dog's stocky frame, the boy was too heavy for him.

Jack was near death. The harsh blow delivered by the horse had resulted in a severe skull fracture, and he would die without immediate attention.

Suddenly the bulldog spied Jack's cap, lying several feet away. Always firmly fixed on the boy's head, it was something Jack never let out of his sight. During the horse's attack, however, it had been knocked to the ground. The dog ran to it now and grabbed it up, then disappeared into the woods.

At the back door to the country house, the bulldog scratched furiously, the cap held in his teeth. Mr. Sigmon, Jack's father, was sitting at the kitchen table when he heard the ruckus. He looked out through the mesh in shock at the breathless little dog and Jack's drooping cap.

Mr. Sigmon rose from the table, apprehensive. He pulled open the screen door but the bulldog wouldn't enter. When the man reached down for Jack's cap, the dog trotted just out of reach, his eyes pleading. Raising his voice, Jack's father called to his wife.

They'd never seen the little dog act the way he did then, racing around the yard and clamoring up to the screen in quick spurts. He wouldn't let the cap out of his mouth. His black, close-set eyes watched them every minute. In a flash of understanding, Mrs. Sigmon took her apron off and followed the bulldog out into the yard.

At once the bulldog turned on his short legs and trotted off in the direction of the pasture, with both Mr. and Mrs. Sigmon now at his heels. Still clutching the cap, he led them directly to Jack.

When Jack's parents reached the injured boy, his heartbeat was faint and his breathing shallow. They rushed him to Anadarko, where the doctors didn't express much hope, performing an operation to remove part of his skull. Medical personnel were amazed at the little dog's uncanny ability to get the injured boy help quickly, and credited him with Jack's slim chance for survival.

But Jack Sigmon did not thwart the Boston bulldog's stubborn efforts to save him. Several weeks after his operation, he had recovered enough to join his parents at their home in Cushing, where he outlived the bulldog.

The anonymous bulldog, who saved Jack Sigmon's life, died in 1930 and was buried in an unmarked grave. However, he was remembered with a nomination for a gold medal in the Latham Foundation's search for heroic animals in 1931.

Zorro

On a backpacking trip with his master, Mark Cooper, the stocky, big-footed German shepherd/wolf named Zorro loped easily through the rugged Sierra Nevada foothills east of Sacramento, California. It was a bright but cool day in November, 1976. Twenty-six-year-old Mark, ahead of two friends, was carefully making his way atop a steep trail which offered a grand view of one of the many canyons in the region. Far below him gleamed a sun-struck river.

Suddenly Mark lost his footing, clutched madly at thin air for something to hold on to, and plunged 85 feet down the near-vertical incline into the ravine below. Bouncing off boulders and being scraped by sharp rocks over the long, agonizing plunge, Mark at last halted, face-down in the river. His body was bruised and battered, his insides hemorrhaging and his pelvis broken.

Zorro had been the only one to see his master fall. Immediately the dog scrambled over the cliff edge after Mark, zig-zagging and falling swiftly down the slippery incline. With his eyes on the helpless man, he saw him being tugged in circles by a strong, swirling whirlpool.

Reaching the water's edge, Zorro leaped into the cold river. The furious water roared and smacked around him as he swam through the turbulence. The closer Zorro came to the center of the whirlpool, the more it worked to suck him downward.

Keeping his eyes on the backpack Mark still wore, Zorro swam hastily toward his drowning master. At last he reached him and closed his teeth around the nylon bag.

Kicking out, the shepherd/wolf began tugging Mark to shore. Just as mightily, the current yanked the man from his grasp. Battling the water, Zorro turned back and grabbed for Mark again. Once more he dragged his master toward shore, straining against the mighty suction until he had him safely away from the whirlpool's grasp.

As he neared the bank, Zorro scrambled along the rocky riverbed with his burden in tow. As gently as he could, he eased Mark from the shallows and onto the river bank. Then the soaked and exhausted dog raised his nose toward the sky and howled for help.

At the top of the trail, Mark's backpacking companions heard the insistent cry and used it to locate the pair. Looking down into the forbidding ravine, they quickly decided the incline was too steep for them to attempt a rescue. Calling down to the crushed man that they were going for help and would return as soon as possible, Mark's companions disappeared from sight.

Zorro, watching them set out, barked a reminder until they were out of earshot. All too soon the sunlight slipped behind the mountain's rim, and the long, cold shadows crept over the huddled man and his dog.

Zorro whined and cautiously lay down across his shivering master's body, his own body heat protecting Mark from the deadly threat of hypothermia. Burying his fingers into Zorro's warm fur, Mark hung on for his life. It would be a long night.

As the temperature dropped in the canyon, Zorro stayed planted over his often-delirious master. Alert to every crackle of branches and skittering of loose shale, the big dog never left the injured man, remaining protectively draped over him throughout the night.

Finally dawn broke over the canyon's east wall. Soon after, a team of rescue workers air-lifted Mark from the narrow ravine by helicopter and, as the ground crew left the area in a single crowded jeep, they called to Zorro to follow them out. Zorro ran along beside the jeep for a while, but sometime during the day he left the rescuers and circled devotedly back to the river, the last place he'd seen Mark. That's where the shepherd/wolf was himself rescued several days later by members of the Sierra Club. Hungry and spent, he was lying by his master's deserted backpack.

After a long stay in the hospital, Mark Cooper recovered from his injuries and was reunited with the dog that saved his life. For his brave and skillful rescue, Zorro was awarded the Ken-L Ration Dog Hero of the Year Award, and the Paws for Love Award for heroism presented by Feline and Canine Friends, Incorporated.

Dumb-Dumb

The normally friendly, silky black cat, Dumb-Dumb, shared garage space with Che-Che, the family's Chihuahua, so both of them heard the prowler at the same time. Che-Che barely raised her head at the stranger's intrusion into the garage, then settled back on her dainty paws with indifference or timidity. But Dumb-Dumb, the five-year-old half-Siamese, bristled immediately.

Dumb-Dumb was a pet of the Russel Carter family of South Wichita, Kansas. That day in 1972 she watched the man carefully as he reached for the door leading from the garage into the kitchen where Mrs. Carter worked, unaware of the intruder. The burglar was keeping a wary eye on Che-Che despite the dog's neutrality, but he paid no attention to Dumb-Dumb, crouched on a shelf behind him as he turned the doorknob quietly.

Dumb-Dumb silently hunkered down on her shelf, gathered her tensed muscles beneath her, and rocked quickly from side to side to calculate the distance of her leap. The burglar had just gotten the door ajar when she sprang.

The cat landed with a tenacious clawhold on the man's head. Now yowling, she clung to her perch with ferocious tenacity. The burglar, surprised and in pain from her sharp claws, whirled in alarm, colliding with shelves and scattering their contents.

Dumb-Dumb was cursed and pounded. Then, with a yank and heave that cost him dearly in torn flesh, the burglar succeeded in tearing her free and flinging her to the ground. At that moment, Mrs. Carter, alarmed by the commotion, opened the door.

With renewed rage, Dumb-Dumb sprang once more to the intruder's back. This time the valiant watchcat got a better hold. Her foreclaws sunk deep into the flesh of his neck and cheeks, her back claws pummeled and scratched incessantly, and she reached boldly around to clamp her teeth in his ear. The man screamed and flung himself through the garage, but was unable to dislodge her. Dumb-Dumb continued the attack and the man bolted from the garage and fled the property.

To the relief of Mrs. Carter, Dumb-Dumb returned after awhile, unharmed and as mild-mannered as ever.

Dumb-Dumb's courage would earn her the Carnation Company's national Cat Hero Award for 1972, her own engraved silver feeding bowl, and a sign posted as a warning to other would-be intruders: "Beware of Dumb-Dumb."

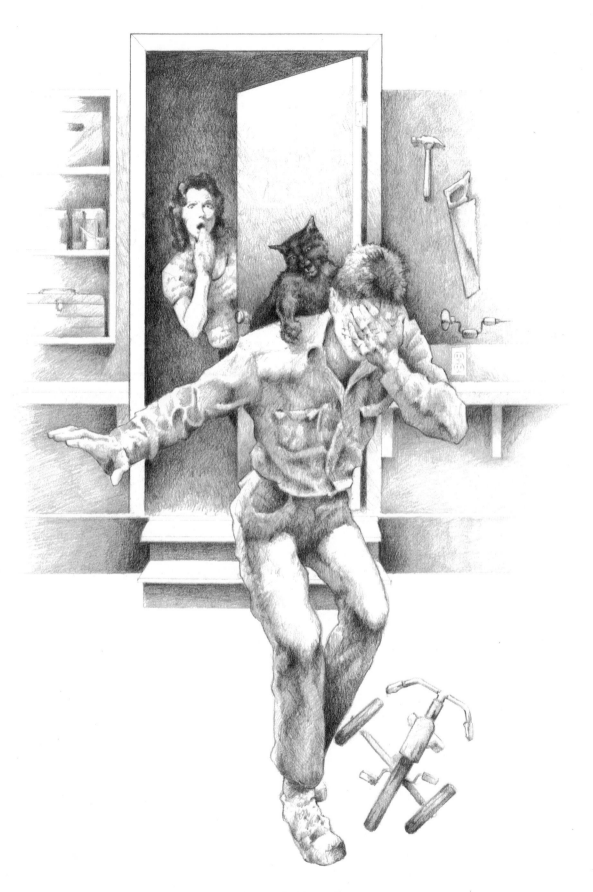

Dumb-Dumb

Prince Pluto

The riptides were strong and the surf was heavy that gray day on Newport Beach, California, as Prince Pluto followed his master, Richard "Shorty" Gunther, to the pier. It was Pluto's morning ritual to tag along with his pal until the fishing boats went out to sea. Then the shaggy, 150-pound Saint Bernard would make his rounds of the beach and nearby towns.

The gentle dog, never trained for lifesaving, had become the self-appointed lifeguard for bathers in distress on the long, notoriously dangerous ocean stretch, and four times already witnesses had watched him save children from the crashing breakers. The big, trundling rescuer seemingly had no trouble distinguishing between a child at play and a child in trouble, streaking into the foamy waves after helpless victims whenever the need arose.

In retrospect, it seems an ironic touch that Pluto, no stranger to rescue, had once been saved himself. When he was a pup, a kindly veterinarian had taken pity on him after his original owner gave him up as a hopeless chicken chaser and ordered him put to sleep. The veterinarian, seeing something special in the mischievous dog's eyes, had laid his hypodermic needle aside and found him a home with Shorty.

They were still together now, four years later, as day broke that chill morning in 1931. Pluto said goodbye to his companion at the pier and trotted off.

For a while Pluto watched six-year-old George Mades of Brea, California, romp in the waves while George's father stood nearby, engrossed in surf-fishing. Although the sky had remained overcast and the tall breakers beat an ominous rhythm, the little boy ran gamely in and out of the shallows. He delighted in daring the surf to "catch him."

Only Pluto saw that playful dare become a reality.

One minute George was standing on firm sand, and the next a series of large, fast waves swept in to roll over him. The little boy opened his mouth to cry out but the water filled his mouth and throat, smothering his terrified shouts. As the tidal surge rushed back to sea, he was tumbled head over heels, smashed downward to the ocean floor, and then shot up like a piece of flotsam. Within seconds he found himself yards out in deeper water.

Pluto saw the boy feebly resisting the fierce waves and bounded toward the water in a spray of sand. Eyes riveted on the spot where George was fast disappearing, he plunged into the pounding surf.

Kicking out with powerful strokes, Pluto saw George's head bobbing just ahead of the largest breakers. The small child, caught in a swift retreat of waves, was being drawn farther and farther from land and down the shoreline. Already his face was a mere speck in the foam.

147

Pluto fought the rough, conflicting currents as breaker after breaker crashed over him. Again and again he surfaced, choking, and turned back toward George, but the little boy seemed no closer. A strong swimmer, Pluto nonetheless began to tire; the constant battle with the churning ocean taking its toll.

Finally his heroic efforts began to pay off. Slicing through a huge breaker, Pluto coughed out a lungfull of water, caught sight of George again and, with several mighty heaves forward through the surf, began closing the gap between them.

When he reached George, the boy was choking and flailing about madly, only seconds from sinking below the surface. The terrified boy lunged out in a desperate attempt to climb onto Pluto's back, but Pluto resisted, swimming around him in a circle and latching onto the youngster's swimming trunks. Then, cautiously, Pluto headed back to land.

Gasping, the big dog swam with the current, moving down the beach but toward shore, towing George in his powerful jaws. Then, as George seemed to relax, Pluto allowed the exhausted boy to climb onto his back. The Saint Bernard's strength was spent from the added weight and the constant fight with the powerful surf, but he was able to maneuver with the incoming wash of breakers so that each kick of his legs brought them closer to the beach, even as the undertow fought to pull them out again.

At long last Pluto's feet found firm sand and he pulled George, shivering but alive, from the sea. The little boy's grip, strengthened by terror, was so strong that his father had to peel his fingers from the dog's fur.

For his fifth rescue of a drowning child, Prince Pluto was named "Official Lifeguard of Newport Beach" by the town's city council. From then on, he wore a department badge and a safety light attached to a harness for protection from cars during his nightly rounds. Newspapers and newsreels headlined his story and, on the last day of Be Kind to Animals Week in 1931, a large, appreciative crowd watched the great Saint Bernard receive the Latham Foundation's Gold Medal Award for animal heroism.

Top

When they stood side by side, Top's head reached the shoulder of the little girl and his back was level with her waist. In fact, he could have put his great paws on her shoulders and washed her face with his tongue. Instead, the Great Dane stepped lightly, even politely, down the Los Angeles street beside her, the 11-year-old holding tightly to his leash.

Top belonged to the German-immigrant actor, Axel Patzwaldt, but with the children in their apartment complex, it was the dog who stole the show. Lanky and amiable, with a neck big enough for a calf's collar and a clownish face, he was always being asked out to play. On this April day in 1969, the neighbor girl had begged to take him for his walk, and now they were just starting what should have been a nice romp in the California sun.

They reached the first corner. Top's companion stepped off the curb. She was talking to him, petting him eagerly, her small hand resting on his shoulder and her eyes on her feet. Usually she was very careful, always looking both ways, but not this time. She just went on chattering excitedly, moving out into the street.

Suddenly the Great Dane heard a high-pitched roar arrowing toward them. Almost in the same instant he saw, out of the corner of his eye, a glimpse of huge, chrome-plated grillwork and speeding wheels, a shocked face behind the dash.

It was too late for the delivery truck to stop. Top gave a deep warning bark and, in one move, threw himself in front of the child, pushing her backwards hard with his shoulder. There was just time for her to go flying up onto the sidewalk, unhurt, before the speeding vehicle bore down on the place where they'd stood.

The truck's bumper hurtled into Top's backside with sickening force, shattering his right hip and spinning him completely around. Crippled and in pain, the big dog crumpled onto the asphalt.

Now, it was another fine day, and Top scratched at the back door to Alex's apartment, anxious to be let out. The door led to the lawn and pool shared by residents of the apartment complex and the Great Dane, just out of his cast, longed for the warm, early-summer air.

Top had spent seven weeks with his leg in plaster after Alex rushed him to a nearby animal hospital on the day the Great Dane saved his young neighbor's life. It had been a rough recovery. The dog's leg was so badly damaged he'd limp for life. And now, as he hobbled outside, the exaggerated swagger made him seem even more clownish than before.

Slowly Top limped onto the deserted lawn, sniffing the grass and shaking the kinks from his stiff body. He circled, leaving the grass for the cement skirting that edged the pool. His nose was up now, his high ears alert. Then he looked into the pool's water.

There, two-year-old Christopher Conley, another of the apartment complex's youngsters, was splayed on the bottom of the pool, apparently lifeless, under six feet of water. Great Danes are notorious water-haters, but Top didn't hesitate. He leaped into the pool, bad leg and all, kicking downward to the submerged boy.

Frantically, the dog grabbed at the child's clothing and limp arms, trying to pull him up. But, for all his size and strength, Top was frustratingly ungainly in the pool. Again and again Christopher floated maddeningly from his grasp.

It was no good. Top's lungs pounded from the pressure, the last of his air swirling away in a stream of bubbles. Blowing hard, he surfaced and paddled desperately toward the side of the pool.

Heaving himself up, Top struggled as his big paws slid on the soaked cement beneath him, his weak back leg buckling. Then he slipped and scrambled toward the back door of Alex's apartment, baying with all his might. Behind him the precious seconds of Christopher's life were ebbing away.

Top howled and howled, his deep yowling throaty with anguish for the drowning boy. It was only moments before his fervent baying brought Alex out. Anxiously, Top seized his master's arm and led him to the pool.

A former lifeguard, Alex retrieved Christopher in moments and immediately started mouth-to-mouth resuscitation while alerted neighbors called the rescue squad. By the time paramedics arrived, Christopher was showing signs of revival. Eight hours after being rushed to Citizens' Emergency Hospital in West Hollywood, the little boy was pronounced out of danger.

In less than nine weeks, Top had saved two lives. For his bravery and intelligence in rescuing both his 11-year-old friend and two-year-old Christopher Conley, the lovable Great Dane was awarded the Ken-L Ration Dog Hero of the Year Award for 1969.

Budweiser

Squat-faced and wearing a patchwork coat, 14-month-old Budweiser had been with Nelle and Mr. Carter only four months. When he'd come to them earlier in the year of 1973, he'd arrived with the reputation of being mean and a "biter." The Saint Bernard had spent the first 10 months of his life chained in a yard.

The kind-hearted Carters, however, had given him the run of their John's Island home in South Carolina. Soon the big dog changed and became a playmate for the entire household, being especially gentle with the "senior" pet, a Chihuahua aptly named Tiny.

On this particular evening, the Carters' six grandchildren were spending the night. Mr. Carter had gone to work at the family's restaurant in North Charleston, and Nelle Carter, with her young houseguests – ages four through seven – were alone in the house with Budweiser and Tiny. Budweiser lounged on the front porch, within earshot of the sounds of Nelle bedding the children down for the night.

The explosion was sudden and terrifying! The big Saint Bernard sprang to his feet as the concussion echoed painfully in his ears. Before he could move, a second explosion ripped the air. Fire burst through the house and the powerful suck of flames greedy for oxygen roared through the rooms.

For Budweiser, safety lay a few steps away. All he had to do was bound over the porch rail. But, without hesitation, he heaved his big shoulders against the unlatched screen door, muscling and squirming his way into the flaming house.

When he entered, thick black smoke, as deadly as the fire itself, was already billowing. Flames crackled on the ceiling, and the floor was hot beneath the sensitive pads of his feet. In the chaos, he could hear Nelle's strained voice calling to the crying children as she tried to gather them together.

He moved fast, bounding into the room where the youngest child, four-year-old Linda Lawson, stood paralyzed and whimpering in shock. Padding to her side, he gently closed his teeth on her shoulder as carefully as if she were a baby chick. The message was gentle but firm as he tugged her toward safety.

Moving fast, never loosening his grip, he guided Linda across the hot floor and out of the flames, not stopping until he'd reached the safety of a neighbor's porch. Linda, frightened but safe, stayed where she was, watching the big dog turn back toward the burning building.

Smoke clouds billowed from the roof and the heat was searing. Through the roar Budweiser heard Nelle's voice, ever more frantic. She had noticed him guide Linda to safety, and she had four of the other children at her side, but she couldn't find five-year-old Joyce Hinson. Frightened and confused, Joyce was stumbling through the smoky rooms, crying and calling out in terror.

Budweiser charged back into the burning house, heading straight for Joyce. In the few minutes since the explosion his hair had been blackened, his feet burned and his lungs seared by the hot smoke.

Once at the child's side, Budweiser took her sleeve in his teeth. Then, as fast as she could follow, he pulled her from the inferno.

Nelle and the other children had gotten out just ahead of them. As she and the children gathered together, Budweiser looked at those safe outside the collapsing house and whined urgently. His little friend, Tiny the Chihuahua, was still inside.

Barking desperately, the big dog ran again into the flames. They were everywhere now, hissing and crackling, curling from the ceiling and devouring the air. They formed a solid, impenetrable sheet before him. Budweiser had saved the children, but he couldn't help Tiny.

An instant after he'd withdrawn from the inferno, the roof fell in where he'd stood. On scorched paws and shaky from lack of oxygen, he limped back to the solace of his family.

Budweiser was treated for smoke inhalation and the burns on his paws, while Mrs. Carter and the six children were uninjured. Tiny was the sole casualty from the explosion and fire.

In recognition of his loyalty and courage, the valiant Saint Bernard was presented with the Ken-L Ration Dog Hero Award for the Year 1973.

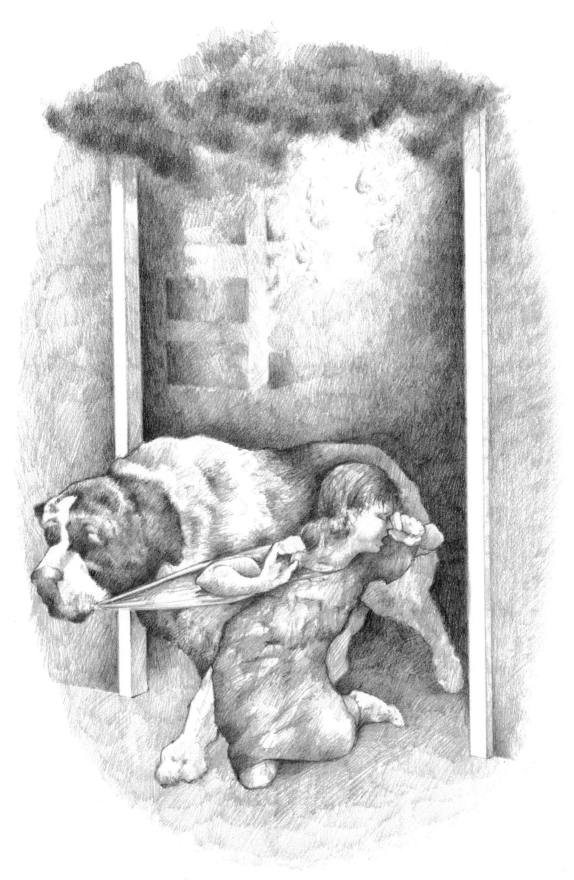

Budweiser

King

II

The knock on the door was loud and abrupt. King's keen German shepherd ears pricked forward at the sound. At nine years old, he was heavy-shouldered and rangy, with a tawny coat and big brown eyes that gleamed with affection for his 77-year-old master, Thomas Perkins.

It was a cold January night in Boston, Massachussetts, in 1986. King had been taken in by Thomas four years before, when his former owners had moved away. Thomas had paid $65 for him.

Now, as the old man shuffled toward the door, King remained alert as always to the sounds within the apartment that had become home to him. The knocking grew more insistent and he saw Thomas reach for the latch.

Suddenly the door swung back on its hinges with a loud crack. What happened next seemed to happen in slow motion. King's beloved Thomas stepped back in fright from the intruder, who moved toward him, shouting. King could see Thomas' eyes widening; could smell his fear as the threatening voice filled every corner of the room.

"Give it up, old man... I know you've got your Social Security check... give it up! Don't run... don't move or I'll blow you away!"

The intruder made a sudden move, and when he drew his hand into view, there was a gun in his fist. Something was terribly wrong. Thomas was running. Running. Thomas, the man who fed him, fondled his ears, spoke in that cadence King cherished as unique to him alone, was running away in panic.

Something within King snapped. A low growl came from his throat. His lip curled back. The blood surged in his veins. He gathered himself up and, in one powerful spring, lunged at the man with the gun, sending him sprawling.

From all indications, King's initial attack was meant simply as a warning. His intentions were to hurl this man back out the door, away from Thomas. Now the intruder would know to leave them alone; that he wasn't welcome in their home.

King waited, his fur raised in warning, while the intruder got up. Then he saw the gun barrel swing toward him. He leaped forward, but it was too late.

The first .22-caliber bullet slammed into King's left front leg. He reeled in mid-air from the pain. Another bullet, in quick succession, stung terribly as it sliced into the bone and cartilage of his paw.

King rolled away, stunned. The intruder kept his weapon trained on him, and there was murder in his eyes.

Instantly, King was on his feet again, eyeing the hated gun. The lethal barrel was just inches from his head. The wounds it had already inflicted oozed blood. His splintered left leg quivered in shock, and his torn right paw hung useless. But King could sense, through the blur of his own pain, that Thomas was nearby, huddling in the other room and still in danger.

The big dog crouched, growling and snapping. Then his long body arced up in a single movement from the floor. His own powerful weapon, his teeth, went for flesh.

In the endless moments that followed, King and the intruder were locked in a violent embrace. The intruder's hot breath, so near, flattened his fur. Frantic arms flailed the gun repeatedly toward him.

Another shot rang out, and another. King felt the now-familiar pain, like red-hot needles, scorching his neck, once, twice. His jaws closed around the arm that held the gun.

Now King was bleeding profusely, but he wouldn't give up. He was 98 pounds of fur, flesh and fury, fighting for his life and that of his master.

It was a vicious struggle, but when it was over, King had won. He drove the now-frightened and mauled intruder out a window. Only then, with his energy spent and the pain of his wounds overpowering him, did King fall to the floor in an exhausted, trembling heap.

Thomas, who'd hidden himself in a closet during the battle, came out unscathed. When he bent over his long-time friend, he found the dog's tawny coat matted with blood, his breath ragged and shallow, and his quick brown eyes glazing over. Despite his age, the old man scooped the dog up in his arms and went for help.

King was taken to Angel Memorial Hospital in Boston. The next day, his heroic actions were recorded in newspapers across the country. By facing the intruder's gun, he'd lost a toe on his right front paw, and would always carry two .22-caliber bullets lodged, harmlessly now, in the muscle sheath around his neck. But he'd survived – and he had saved Thomas.

Shotgun

Rain and gale-force winds lashed the tug *Barney Jr.* as it floundered in heavy seas off Angeles Point, located in the Strait of Juan de Fuca eight miles west of Port Angeles, Washington.

Caught aboard in the squall that May 16, 1930, were the ship's cook, C.H. Coulson, and the wireless operator, Arthur Clayton. Clayton tapped out one last desperate S.O.S. as the squat tug lurched leeward and rolled sickeningly. Hearing Coulson's shout, he pulled his way along the tilted cabin wall to the door. He knew there was no time left — they were going down.

On shore, brown hide drenched and mane dripping, the saddle pony Shotgun twisted her neck for a sidelong glance at her master, Albert Smith. He cinched the girth a notch tighter and swung onto her back. It was dangerous weather to be out in and Smith, sensing her apprehension, gave her a hasty pat on the withers, shouting instructions to her above the wind. Then, slicker tenting out and hat pulled low, he spurred the stout little mare through the gates of his resort, "The Place," on the lower Elwah River. Together they raced along the rain-pelted estuary toward the churning surf.

Galloping along the beach, Smith could no longer see the wallowing *Barney Jr.* that he'd spotted from a rise above his resort. Scanning the broken sea he spotted the boat at last. She'd upended about a quarter of a mile out, like a discarded toy. The tiny figures of the men clinging to her deck waved their arms in a frantic plea for help, their shouts carried away by the crash of the storm's monstrous swells. They couldn't swim for it. Given the 15-foot waves and perilous undertow, there was little chance they'd make it to shore.

Smith bent forward to shout in Shotgun's ear. "We're going in!" he cried over the roar of the wind. Giving rein, he urged the saddle pony forward.

Shotgun, obeying Smith's commands, galloped along the water's edge and plunged into the foaming surf. At once a fierce cold sucked at horse and rider. Salty water stung their eyes with the force of buckshot. Ahead of them lay a vast expanse of dark and seething sea.

Shotgun's legs were short and powerful, yet the tiny horse was no real match for the surging water. Quickly the veins along her short neck stood out and her eyes bulged with effort. For as long as they could they followed a sandy ledge. Then the ledge gave way and Shotgun ploughed dutifully toward the sinking tug. She'd been battling the sea for a good 15 minutes.

Coulson and Clayton peered incredulously at the snorting horse and drenched rider appearing from nowhere through the rain. Experienced seamen, they knew all too well there was little hope of a man making it through that treacherous surf. But a horse?

"Lower the lifeboat!" Smith shouted up to them. "She'll pull you in." The tug tilted dangerously and the two men scrambled to comply. Crawling starboard they released their dinghy, struggled into it and rigged an improvised lifeline, tying one end to the bow and tossing the other to Smith.

Overhead the squall shifted angrily and pummeled them with renewed force. The dinghy, barely big enough for the two men, bobbed toward the open sea. Quickly Smith grasped the line in near-frozen fingers and lashed it to Shotgun's saddle horn.

Responding to Smith's signal, Shotgun gave a low whinny and heaved herself toward shore. A few yards from the *Barney Jr.* they felt a great backward surge as the stricken ship went belly up. Shotgun, straining mightily for solid ground, did not look back.

The horse was caught in the quick rolling crash of the waves. Water surged from right and left as she fought to steady her course, all the while tugging the added weight of the dinghy and its human cargo. At last the pony felt the sandy ledge under her hooves. Fighting against the heavy drag of the burden behind her, she took one stubborn step, and then another. They'd completed the most precarious leg of their mission, passing through the breakwaters.

Suddenly Shotgun took a step and dropped from sight, disappearing into a giant cleft made by the swirling tide. At the same moment a comber knocked Smith from the saddle and swept him away. Now, without her master to guide her, Shotgun kicked frantically under the dark, swirling sea, trying to regain a foothold. The lifeline still tied to the empty saddle hampered her efforts, and the tonnage of crashing waves pressed her down. Rocked by the huge breakers, the men in the lifeboat were both helpless to abandon it or to do anything for Smith. Briefly they caught a glimpse of him fighting for his life in the immense waves off to their right. Ahead, through the wall of water, they made out the half-submerged Shotgun. The sea dashed over her again and again as she wallowed in the grip of the whirlpool.

At last they felt a tightening of the lifeline once more and peered over the bow to see Shotgun straining and lunging unguided toward shore.

Shotgun had her footing now. All along the ledge she shouldered aside the brunt of the waves. At last the water was only knee high; then only a swirl around her hooves. Shotgun snorted and pulled free of the sea.

Still she didn't stop. Purposefully she pulled the dinghy the last few feet through the surf until the boat was beached. Only then did she look back, large eyes blinking in the rain, staring toward the waves where her master had disappeared.

Fortunately, Albert Smith had not drowned in the stormy gale off Port Angeles. After an anxious struggle, he too made solid ground and, exhausted, dragged himself onto the sand not far from the other survivors. All three men were safe.

For her part in saving the lives of the shipwrecked sailors, Shotgun later received Washington State's 1931 nomination for the Latham Foundation Gold Medal Award. Her courage and incredible stamina were bannered in headlines of the day.

Nemo

Nemo moved like contained electrical energy, powerful and alert. He watched as the young man in combat fatigues approached the wire enclosure and knelt down, speaking to him through the webbing. Then the man, 22-year-old Airman Second Class Robert Thorneburg of the 377th Police Squadron, stood and slipped into the enclosure, approaching his partner of four months with care.

Respect and acceptance were things you earned with Nemo. He had begun his training at Lackland Air Force Base in Texas when he was two years old. The German shepherd was now in peak condition at the age of four and, training completed, was on sentry duty at Tan Son Nhut Air Base near Saigon, Vietnam. His original handler had returned stateside in the fall. Thorneburg, who now began the ritual of checking his dog for sores and scratches, had yet to test the true mettle of their partnership.

Completing his assessment of his partner's condition, Thorneburg muzzled Nemo, snapping on a lead leash. As evening descended on December 4, 1966, the pair fell into formation and climbed into the back of a military truck with several other sentry-dog teams. This night, as on every night, there would be a new posting and a different drop time.

In marked contrast to other nights, this particular night's G.I. banter was quiet and subdued. Sensing the change, Nemo sat rigid as the truck lurched forward on its rounds to drop its load of sentry teams on station.

Police Squadron partners Nemo and Thorneburg were a highly trained team. They had to be. With the other teams in their squadron, they spent each night alone patrolling the perimeter of the vital air base. Their job during the tense dangerous hours of darkness was to spot and challenge any infiltrating Viet Cong, who had become increasingly determined to destroy the base's aircraft and facilities.

If warned of an enemy's presence by Nemo's sensitive nose, Thorneburg's orders were to radio their position to Central Security Control (CSC), where the duty officer, map plotter and communicator stood by. Then, with Nemo in the lead, they'd move in and investigate. If Nemo's warning proved positive, Thorneburg radioed confirmation and security alert teams, armed with M-16 rifles, machineguns, grenade launchers and flares, were called into action.

A lot could happen in those precious moments before help arrived. The previous night's predawn attack had proven that. In one of the first battles of its kind, the 377th had, for seven bitter hours, effectively flushed out and held off a superior VC force intent on assaulting the base. Although dawn found the base secure, three airmen and their dogs died in the fighting.

This evening, everyone was alert – and edgy. Thorneburg and Nemo had pulled duty near an old Vietnamese graveyard, which lay within the perimeter of the base. The sound of the truck's engine disappeared into the heavy darkness. Side by side they began their patrol.

Nemo paused beside a shadowy Vietnamese shrine, tensing in the cemetery's silence. Beside him his handler cast a quick glance at the big dog's bright eyes, barely discernible in the light from the stars overhead. The shepherd's ears perked and his ruff bristled. Man and dog froze. A stiff breeze rustled the tall, thick elephant grass growing along the edge of the graves.

Suddenly, Nemo whirled and glowered in the direction of the elephant grass, a growl stifled deep in his throat. Silently he coiled into a ready stance at the end of his leash. Beside him, Thorneburg crouched and quietly released the clasp on the leash, freeing Nemo for action – action which would give Thorneburg the precious seconds necessary to switch on his radio and alert backup forces.

Nemo charged and the elephant grass closed over him. In the deeper darkness he raced toward the enemy, overtaking them within moments. Four Viet Cong from an advanced army unit had slipped through the outer perimeter and were making stealthy progress toward the base's vulnerable aircraft.

Without hesitation Nemo engaged the enemy. The VC soldier to his right, hearing the dog's rushing charge, swung his automatic rifle up in desperation. A burst of fire shattered the ground under Nemo as he leaped. The guerrilla fell, with Nemo at his throat. He died almost immediately from the dog's powerful attack.

Having alerted the base, Thorneburg raced into the underbrush thinking Nemo had flushed out a lone sniper. He suddenly came face to face with the three remaining VC. Now trapped, Thorneburg's sole advantage consisted of the precious seconds Nemo had bought him with his surprise attack. Thorneburg charged the nearest VC soldier and began fighting for his life.

In that same instant Nemo threw himself between Thorneburg and the remaining VC. This time the enemy was less surprised. The breath left Nemo as a VC's rifle butt caught him in the chest. Knocked sideways, Nemo slammed painfully into the damp earth but rallied instantly, springing into battle again.

Now he had his man. They crashed through the underbrush, locked in mortal combat. Their eyes glittered inches apart. The VC maneuvered his automatic rifle into position, pointed directly into Nemo's muzzle – and fired.

The impact of the bullet sent Nemo crashing backward. Flames and hot metal smashed into his jaw and split upward, driving through his eye socket. Bone splintered like shrapnel, and blood gushed from the wound, blinding him. Another blast from the automatic cut the air above Nemo's head.

Struggling to stand, he shook the blood covering his one remaining eye and saw the weapon now aimed at Thorneburg. Bullets tore into his handler's shoulder even as Nemo lunged forward with a single sound in his throat. It was enough to distract the VC, forcing him to turn his weapon toward the renewed threat. The rifle barrel which had just torn Nemo's face apart was leveled at him again. Nemo charged straight into it.

Nemo

In a high arc, Nemo met the VC soldier head on. Bullets sliced uselessly into the jungle as the dog gripped his enemy's arm with shattered jaws, tearing and ripping until the weapon fell. Then swiftly Nemo killed his enemy.

Thorneburg lay wounded on the ground. As help arrived, his courageous and determined canine partner, with a bloodied eye hanging useless from its socket and a painfully ripped muzzle, launched himself on the remaining VC.

The skirmish ended in a volley of fire. Bullets sailed through the elephant grass, both from friendly and enemy forces. The badly injured dog, with draining strength, held his position until the defending airmen had cleared and secured the area. Thorneburg and Nemo were evacuated, seriously injured but alive.

After the battle Thorneburg was sent to Japan to recover from his wounds, separating him from his partner forever. Nemo spent eight months convalescing in a veterinary hospital.

The sentry team's bravery was not forgotten. When Nemo sounded the alarm that December night, he not only saved his partner's life, he alerted the entire base to the threat of the VC attack. His intrepid action was recognized in national newspapers and his name became synonymous with the bravery and sacrifice shown by all "guided muzzles," as Vietnam's sentry dogs were known.

Algol delle Aque Celesti
(Blue Water Algol)

Heavy rains lashed western Italy all that February day in 1980. Because of the storm, and his wet, tangled fur, Algol delle Aque Celesti, a two-year-old Newfoundland show dog, had been groomed and bathed professionally, rather than at home, for his important upcoming show.

When his master, Dr. Carlo Orlandi, picked Algol up that evening, the stockily-built dog greeted him with his usual enthusiasm. Fur sleek and shining, big feet in a near prance, the young Newfoundland looked like just what he was – a fast-growing dog barely out of puppyhood. Orlandi smiled at the fine picture the proud Algol made, took hold of his collar, and together the pair made a dash through the rain-swept darkness for the car.

It was after nine o'clock, and the roads out of town were nearly empty. The rising storm, battling at the little car's fogged windows, seemed to have turned its full fury on the two lone occupants making their way along the road leading to Orlandi's home. The wind shrieked from no apparent direction, while leaves, branches and even a small, uprooted tree tore across their path.

Orlandi frowned as he turned off the highway and toward the little railless bridge leading to his villa. Rolling down his window to clear the fogged windshield, he patted Algol, who whined in apprehension. In the headlights the usually tiny creek was swollen clear to the trestles and clogged with debris.

But they were nearly home. Both dog and man could see the lights Mrs. Orlandi had left on for them, burning dimly across the sodden fields. In a few more minutes they'd be safe and dry by the fire. Orlandi drove the car onto the narrow bridge.

Maybe it was in anticipation that the doctor, tired from the day's work and the tensions of driving through the storm, turned too sharply. Maybe it was just the misguiding pull of the wind. Whatever it was, the car's front wheel left the safety of the wood and spun, suddenly useless, in the air. Instantly the car was in space, rolling over to land upside down in nine feet of water.

There was a terrible crack as Orlandi's head hit the steering wheel. Water gushed through the open window at his side while Algol, bruised and shaken, righted himself and nudged the inert man. Orlandi didn't stir. His body shifted crazily in the dark and rapidly sinking tomb.

Algol found his master's arm and clamped his teeth around it, pulling furiously. But the doctor's shoes were tangled in the foot pedals, and he couldn't be budged. Frantically Algol tugged and bit at the twisted cloth. As he worked, the fast-rising water closed over him and soon cut off his air. Finally he could

165

stand it no longer; the near-drowning dog fled through the open window and surfaced into the battering storm.

Paddling furiously against the swift current, Algol filled his lungs in great gulps, and then, with hardly a glance at the shore and the lights of home, he plunged beneath the surface of the dark waters.

Diving back through the car's open window, Algol maneuvered as quickly as he could around Orlandi's still form. He went at the tangled clothing until once more his need for air drove him back to the surface.

Again and again he dove to the car. Soon his big frame shook involuntarily from the pain of his injuries and the exhaustion that was rapidly overtaking him. Each time he swam through the car window to jerk and tug at the clothing that held his master prisoner. He yanked at Orlandi's ankles and pulled so hard at his wrists that the flesh tore.

Was it hours, or only minutes? Now the pain and exhaustion seemed to be lessening. It was as though he'd spent a lifetime in the eerie underwater currents and shadows, and the depths were suddenly beguiling. All he had to do was to forget his burden, stop his mad tug-of-war and float away.

No! There was still a spark in the drowning dog that would not give up. Gasping, Algol surfaced weakly, gulping air and water, then dove one last time for his master.

Orlandi finally broke free. Inch by inch Algol was able to guide him, his mouth clamped firmly on Orlandi's coat collar, toward the car window. Then, at last, his burden slipped out, and he was able to pull his master to the surface.

Algol's body, his eyes and ears and nose, had lost their senses of perception. Even as he surfaced with Orlandi, the storm's great renting seemed to hold little sway over him. He barely heard the wind or felt the whip of rain; hardly saw the light of Mrs. Orlandi's flashlight coming closer.

There remained one small area of the car still jutting above the swirling water. It was here the courageous young Newfoundland dragged the lifeless form of his beloved master, who had been killed instantly in the crash, and collapsed beside him.

In the last flicker of his life, Algol delle Aque Celeste rose up and lay his head on Orlandi's still chest.

Acknowledgement of Resources

American Humane Association presents the William O. Stillman Award
This award was created by Mrs. Morris H. Vandergrift, Philadelphia, Pennsylvania in 1928 for rescue of animals or animals' rescue of humans.

Feline and Canine Friends, Incorporated
Founded by Rose Hosea, headquartered in Anaheim, California, presents the Paws for Love award for animal heroism.

Friskies PetCare Division, Carnation Company
From 1968 through 1984, Carnation conducted (through The Friskies Cat Council) the annual Cat Hero program. The Cat Hero for each year was selected from a field of finalists, nominated from around the country, by a panel of independent judges. The selected Hero was awarded an engraved silver feeding dish, a U.S. Savings Bond and a year's supply of Friskies cat food.

Ken-L Ration Division, Quaker Oats Company
The Dog Hero of the year awards program began in 1954. Each year dogs from around the nation are nominated for the award. From the finalists a panel of judges select the dog to be Ken-L Ration Dog Hero of the year. The award of the gold medal, a $1,000 U.S. Savings Bond, a gold-plated leash and collar and a year's supply of dog food are presented at an annual banquet in honor of the chosen canine hero.

The Latham Foundation
Headquartered in Alameda, California, Latham was founded in 1918 by Edith Latham and her brother Milton. The underlying message was and remains – "respect for all life through education." From 1930 through 1945 the Latham Foundation awarded individually designed medals to selected animals who performed heroic deeds.

Today, besides producing and distributing films, Latham sponsors conferences and seminars, publishes a newsletter, promotes research, facilitates people and pet projects, and is a resource and idea exchange center. Among its many works the Foundation has published *The Loving Bond: Companion Animals in the Helping Professions*, edited by Phil Arkow. This 420 page volume is a comprehensive study guide and resource on animal and human bonding, especially as applied to animal companions for the handicapped and pet therapy. (Available through The Latham Foundation, Latham Plaza Building, Clement & Schiller Streets, Alameda, California 94501. Cost is $19.95.)

Texas Veterinary Medical Association
TVMA founded the Texas Pet Hall of Fame in 1984 to honor animals who, through unselfish and courageous accomplishment, exemplify the human/animal bond.

Actors & Others for Animals, North Hollywood, California.

American Cetacean Society, San Pedro, California, Patty Warhold.

American Kennel Club, New York, New York, Brenda Smith.

American Society for the Prevention of Cruelty to Animals, New York, New York, Margaret M. Holman.

Animal Organizations & Services Directory, edited by Kathleen A. Reece, published by Animal Stories, Huntington Beach, California, $16.95.

(Continued)

Acknowledgement of Resources (Continued)

Argus Archives, New York, New York, Julie Van Ness.

Canine Companions for Independence, Santa Rosa, California, Janet Herring-Sherman.

Liz Cullington, Pittsboro, North Carolina.

Department of Defense Dog Center, Lackland Air Force Base, Texas, Hildegard Brown.

Department of Forestry, Sacramento, California.

Wade Doak, Whangarei, New Zealand.

Dogs for the Deaf, Encino, California.

Dog Sports Magazine, Benicia, California, Mike McGowan.

Fish & Game Department, Sacramento California.

Humane Society of the Pikes Peak Region, Colorado Springs, Colorado, Phil Arkow.

Shirley Keith, Santa Barbara, California.

Ernest Lloyd, Author, *Animal Heroes,* Pacific Press Publishers, 1946.

Massachusetts Society for the Prevention of Cruelty to Animals, (publishers of *Animals* magazine), Boston, Massachusetts. Marilyn Stanton.

National Humane Education Society, Leesburg, Virginia, Anna C. Briggs, President.

National Marine Mammal Lab, Seattle, Washington, Howard Braham.

Pets & Pals, Incorporated, Sacramento, California.

Sam Ridgeway (author of *The Dolphin Doctor,* Yankee Publishing Company, Dublin, New Hampshire, $12.95), Naval Ocean Systems Center, San Diego, California.

Sacramento Pets in Need, Incorporated, Sacramento, California.

Sacramento Society for the Prevention of Cruelty to Animals, Sacramento, California.

San Diego County Humane Society & Society for the Prevention of Cruelty to Animals, San Diego, California, Larry Boersma.

San Diego Wild Animal Park, Escondido, California.

Sea World, San Diego, California.

Signal Corps Museum, Department of the Army Museum, Fort Monmouth, New Jersey.

United States Air Force Historical Research Center, Maxwell Air Force Base, Alabama, Lynn Gamma.

United States Army Military Police School, Fort McClellan, Alabama, Mary Himes.

United States Center of Military History, Washington, D.C., William J. Webb.

United States Marine Corps Historical Center, Washington, D.C., Danny J. Crawford.

United States Police Canine Association, Upper Marlboro, Maryland, Richard O. Rogers.

Sister Maria Veronica, Medal of Honor Archives, Freedoms Foundations, Valley Forge, Pennsylvania.

Western Horseman, Colorado Springs, Colorado, Randy White, Editor.

Wisconsin Department of Natural Resources, Bureau of Endangered Resources, Madison, Wisconsin, Richard Theil.

Wolf Park, North American Wildlife Park Foundation, Battle Ground, Indiana, Erich Klinghammer.

Lotus Teagarden Sturgeon, Los Angeles, California. Thanks Mom.